UNIVERSITY PHYSICS

A Calculus-based Survey of Physics

VOLUME 2

APPLICATIONS OF MECHANICS

BY
MOHAMMAD SAMIULLAH

Published through Createspace
Charleston, S.C., USA

University Physics
Samiullah, Mohammad

Cover Design: Createspace.com.

ISBN-13: 978-1477470183
ISBN-10: 1477470182
Library of Congress Control Number (Vol 1): 2012908078

Typed in LaTeX
Published through Createspace, Charleston, S.C., USA.
Distributed by CreateSpace.com

About the Author

Mohammad Samiullah has taught introductory and advanced physics for 22 years at Truman State University, where he is currently Professor of Physics. Dr. Samiullah earned two Ph.D.'s, one in Chemistry and the other in Physics, from Boston University and was a Research Fellow at the Indiam Institute of Science, Bangalore before joining Truman. He has a Master's degree from the Indian Institute of Technology, Kanpur. Dr. Samiullah has considerable experience with research in Physics Education and has published widely on collaborative learning methods in physics teaching. He was awarded a Truman Fellow and a Jepson Fellow at Truman State that recognized innovative teaching methods. Dr. Samiullah's research interests in physics have spanned from Cosmology to Solid State Physics and Nonlinear Optics.

Preface

A solid foundation in physics is necessary for many disciplines in the pure and applied sciences. This requirement has been traditionally satisfied by a one-year long course in general physics, which covers mostly topics from classical physics. Over the years a number of textbooks have been written to help students and teachers of these courses. These books have also evolved to incorporate the findings on physics education research. In order to make the first year physics more accessible and appealing to students the books have attempted to present the same material in various formats, more colorful graphics, systematic problem solving strategies, online homework access, etc. These changes have come at a considerable cost. The prices of the books have gone up several fold and the content of the books have become less rigorous.

The loss of rigor has been very unfortunate and detrimental for students. For instance, it is well-known that many students now do not develop understanding of basic concepts such as an vectors and kinematics. Many more students do not learn to think physics from a conceptual perspective, instead, they learn to rely on formula-manipulations without understanding where those formulas come from. As a result, even the best students appear unable to derive simple results after successfully completing a one-year course in physics. Often the wide range of the coverage material is blamed for these deficiencies. However, an examination of the current books reveals that they present various topics in essentially unrelated way and do not emphasize enough the few basic ideas which are essential for understanding physics. This type of presentation does not lend itself to teaching students the important of derivations and places undue emphasis on the use of final results in mostly numerical problems.

This book is based on a first-year course that I have taught for a number of years at Truman State University. It has a standard selection of topics suitable for the first-year physics course but differs from the existing books in the degree of emphasis placed on the understanding of the fundamental ideas and the care with which the derivation of various results are presented.

My general approach in writing this book has been to explain each idea as clearly as possible, develop ideas and concepts starting from few basic principles, and illustrate them with relevant examples containing both numerical and symbolic problems. The level of the rigor has been carefully managed so that the book should be accessible to those students who are concurrently enrolled in the Calculus sequence.

By various detailed guided exercises and challenging problems this book helps students develop their own approach to problem solving rather than simply learn to follow a set of rules. Many books try to encourage students to memorize a step-by-step process of solving certain problems. This approach to physics is contrary to physics thinking as it emphasizes "getting" the right answer rather than reasoning correctly from first principles.

Although learning to follow a set recipe can show better performance on tests that test whether students know how to follow those recipes, this approach turns physics into a subject of word problems rather than a subject where a student learns how to examine real situations through simple models. To help students grasp the approach unique to physics, the examples in this book contain a detailed analysis of various ideas, approximations and their limitations. It is hoped that by studying these examples a student will develop a sense of the "physics" reasoning that he/she can bring to bear when confronting new situations.

The four major topics of the classical physics, Mechanics, Thermodynamics, Electrodynamics, and Optics, are spread over six volumes so that each volume would be easy to carry around in a back pack. These six volumes should serve as a strong base for the future work in physics and engineering.

I am sure there are many typographical errors still left in the text despite numerous revisions. Some students, especially Mr. Brian McClain and Mr. Robert Ashcraft, have read earlier drafts of this text and were instrumental in fixing typographical errors. I am very thankful to several students for their feedback in other courses where parts of the present and other volumes were used. I would appreciate hearing about any typos and mistakes in the book so that I can correct them for the next edition. I would also welcome any other observations or remarks you may have regarding the content and approach of this book.

During the writing I had much support from my family and am truly grateful for that. Without their understanding and encouragement I would not have been able to complete this task.

Mohammad Samiullah
Kirksville, Missouri, USA.

TO THE TEACHER

University Physics is organized in six volumes:

1. Fundamentals of Mechanics
2. Applications of Mechanics
3. Thermodynamics
4. Electrostatics and Magnetostatics
5. Electrodynamics
6. Optics

1. Fundamentals of Mechanics

After the introductory chapter which presents the basic approach of physics, vectors are presented in great detail. Every student is encouraged to master the concepts associated with vectors before proceeding any further in the text.

The important topics of kinematics is presented in an approach that emphasizes the general definition before the particular applications. This order is different from the "normal" practice in other textbooks at this level. I believe that the normal order of doing the one-dimensional problems first is counter-productive to student-learning, and may even be harmful. One-dimensional problems de-emphasize the directional aspect of vectors, thereby introducing many long-lasting misconceptions such as the inability to tell the difference between the magnitude and the component of a vector. Many examples of this deficiency can be cited. For instance, students tend to apply the constant acceleration equations even when they are not applicable, e.g., in the simple harmonic motion. Too many problems done early on in one-dimensional constant acceleration has a way of leaving a lasting impression in the minds of students!

Many teachers feel that students would be far better off by first mastering the correct definitions and then using those definitions in applications such as one-dimensional problems. However, the books seem to go the other way. In this book I have spread the topic of kinematics over two chapters - the first contain the fundamental definitions and the second the applications. A teacher may wish to treat the the two chapters as one and lecture by going back and forth between the two chapters as you develop the concepts of displacement, velocity and acceleration.

The subject of kinematics is followed by a long chapter on forces. Various common forces are introduced and studied in the context of the static equilibrium. This is important because students should learn that forces exist independent of acceleration. When students encounter forces first in the context of $\vec{F} = m\vec{a}$, many confuse the $m\vec{a}$ as being a force. It is difficult for students to separate the two sides of the equality as different physical quantities. Studying statics, where you study $\vec{F} = 0$, before tacking the dynamics helps avoid this problem since students already have experience in dealing with forces independent of any motion.

I have introduced the concept of torque in this chapter so that I could discuss the static equilibrium situation more fully. This also helps develop the concept that forces act at different particles of an extended body. For some teachers, this may be an early introduction of torque. The topics of torque are separated enough that a teacher can design his/her course to skip the section(s) on the torque in this chapter and return to this topic after reaching the chapter on rotation.

The chapter on forces introduces students to a multitude of common forces. The distributive nature of forces on the extended bodies are illustrated and emphasized. To help with the concept of force as interaction between bodies I have introduced Newton's third law of motion at this stage before presenting Newton's second law. This chapter also provides an important learning ground for the techniques of the free-body diagrams.

Next, Newton's laws of motion are presented systematically, and the dynamics of single particles and multiparticles are studied extensively. The dynamics is continued in the next chapter using the concept of impulse. The conservation of momentum is presented, both in the context of the third law of motion and then again in the chapter on momentum.

We continue the discussion on Newton's laws of motion by introducing impulse in the next chapter. This chapter shows the emergence of the principle of conservation of momentum. Important applications to collisions and changing mass systems such as rocket is discussed here.

The concept of work is developed from a fundamental perspective. Rather than use Fd to introduce the concept of work, we go directly to the dot product and the infinitesimal work, $dW = \vec{F} \cdot d\vec{r}$. We develop the main ideas of the conservation of energy from the ground up based on the work-energy theorem. By separating the work on multiparticle systems into the work by the internal forces and the external forces a student is encouraged to see how energy conservation can be applied more broadly. Various potential energy formulas are derived from the first principles and the role of the reference energy and reference point in space are made plain to the student.

After the chapter on the energy, there is a rather long chapter on rotation. A teacher should set aside a good deal of time for a thorough treatment of rotation. The main emphasis in rotation is on the angular momentum in the context of fixed-axis rotation. The conservation of angular momentum is discussed here. The rolling motion of an extended body is developed from the mechanics of the individual particles.

To keep the present volume to a manageable length the other topics such as non-inertial frame, gravitation, oscillations, mechanical waves, and fluid mechanics have been presented in the second volume.

2. Applications of Mechanics

This volume continues the discussions on mechanics. This volume and Volume 1 should be thought of as one book on Mechanics.

Volume 2 starts with a discussion of the modification of Newton's second law in a non-inertial frame. This is necessary for a treatment of the Coriolis force. The chapter on Newton's law of universal gravitation includes an extensive discussion on the planetary motion and Kepler's laws. I have laid out the basis for the effective potential energy and discussed the fundamental approach for studying the central force problem. The orbit equation is discussed in considerable detail.

The two chapters on the vibrations and waves introduce students to the basic aspects of the Simple Harmonic Motion, coupled oscillators and waves. The concept of normal modes are introduced and serve as the basis for a discussion of other topics. I have tried to take the mystery out of the wave motion. Learning the basic aspects of coupled motion helps a student extend his/her understanding to the vibrations in a system where the motion of a large number of particles are coupled.

A chapter on statics introduces internal stress and strain in a solid body. This chapter would be a review for students who have completed the chapter on force presented in the first volume. Finally, the fluid statics and dynamics are presented over two chapters, where the differences between the dynamic pressure and the static pressure are highlighted and clarified.

3. Thermodynamics

The subject of thermodynamics is presented here from the traditional perspective. First, the effects of heat on substances is described in significant detail. The only magic that is thrust on the students without any explanations is the ideal gas law. This shortcoming is remedied in the chapter on the kinetic theory. The ideal gas provides a good system in which various aspects of thermodynamics can be explored more fully by a student. the first chapter of this volume explains students the meaning of heat and carefully defines the concept of temperature based on the thermal equilibrium. The wrong definition of temperature in terms of the kinetic energy of molecules, quoted often in most textbooks, is avoided here. Instead, we rely on the notion of the thermal equilibrium to introduce temperature.

Next we discuss the first law of thermodynamics in an expanded view of the conservation of energy. Various applications of the first law, including Calorimetry, are presented to solidify the understanding of the special nature of thermal interaction between a system and the surroundings. The second law of thermodynamics is next presented in the classic forms of Kelvin-Planck and Clausius, and illustrated by application to the Carnot engine and other practical engines. Clausius's theorem is also proved in this chapter. An entire chapter is then devoted to the concept of Entropy where the concept of reversible and irreversible processes are introduced. Finally, there is a chapter on the kinetic theory where simple models of gases are discussed in detail.

4. Electrostatics and Magnetostatics

I have separated the electrostatics and magnetostatics from a discussion of the electrodynamics because the later in many ways contradicts some of the rules learned in the steady current conditions. The topics in electrostatics and magnetostatics follow the standard topics in the usual order except that I have gone into a little more depth in discussing the more abstract laws - the Gauss's law for electric field and the Ampere's law for the magnetic field. One difference from the traditional books at this level is the emphasis of the field point as an arbitrary point in space. The electric and magnetic field has been labeled with subscript P to continually reinforce that the field concept is a local concept. The vector natures of electric and magnetic fields are reinforced throughout the text and problems.

5. Electrodynamics

The importance of dynamics in the Faraday law is emphasized. Students have difficulty grasping the electrodynamics, as occurring only when the fields are changing and stop when the fields become static. The failure of the Kirchoff's loop rule in a dynamical situation is addressed. The treatment of loops placed in a magnetic field in most textbooks contain conceptual errors. The correct way requires the application of Faraday law consistently when magnetic flux is changing through the loop. A chapter on Maxwell's equations is presented to complete the laws of electricity and magnetism. AC circuits methods have been included as an application of electrodynamics.

5. Optics

The volume on optics is self-contained containing introductions to both geometric and physical optics. Some of the concepts of waves from Volume 1 has been presented once again when discussing the wave optics. A repeat of these ideas in the context of light often gives students a second exposure which is often necessary to learn them well.

This volume begins with a discussion of the fundamental aspects of light. The next two chapters are devoted to geometric optics. The second chapter is entirely concerned with the image formation by reflection and refraction. Various basic optical elements are discussed here. The applications of geometric optics to instruments are presented next, where a detail picture of optics of the eye is also given.

The rest of the book is devoted to the wave or physical optics. In physical optics I have tried to reinforce everywhere the concept of the addition of amplitudes which is central to the understanding of wave phenomena. Some of the sections in these chapters are mathematically demanding, but it would be worthwhile for a student to work though the calculations in the book and attempt the more challenging problems found in the exercises and problems.

TO THE STUDENT

This book was written to provide you with a solid foundation of introductory physics. The book is expected to be accessible to students who are at least concurrently enrolled in a Calculus course. The emphasis in this book is on generating a deeper understanding of physics rather than creating a superficial familiarity with the subject. You cannot gain a good understanding of physics without a lot of effort on your part.

Dear student, you should realize that reading physics textbook requires time and patience. Rushing through the textbook will not be enough to learn physics. You are expected to sit down with the book and really study the book to get the material presented here. To help you take notes down while you are studying I have left a healthy margin on the side of each page. The fonts have been chosen to make the reading less straining to the eye. You would realize that the more time you put into your studies, the more you would get out of your studies, and you will derive much satisfaction from learning the subject.

The chapters are built on each other. You should not skip over any problems you encounter in the early chapters since those problems are more likely to show up again in the later chapters.

The exercises and problems are key parts of the textbook. The exercises and problems at the end of chapters have been selected carefully to improve your understanding and challenge you as well. The exercises should help you improve your understanding of the basic definitions of single concepts, and problems should help you think about multiple concepts together.

Many of the problems are non-trivial and you are expected to spend a good deal of time solving them. These problems would prepare you well for the advanced courses in your major and lead to a fruitful career in the pure and applied sciences. Some of the exercises have multiple parts so that you can explore a physical situation more fully and develop a sense of a more complete understanding.

This book is dedicated to the memory of my father Mohammad Jasimuddin Ahmad, who was also my first math teacher.

TABLE OF CONTENTS
VOLUME 1

TABLE OF CONTENTS
VOLUME 2

VOLUME 2

APPLICATIONS OF MECHANICS

Chapter 1
NONINERTIAL FRAMES

Contents

So far we have studied motion with respect to inertial observers. For these observers, Newton's second law of motion takes the familiar form, "the rate of change of momentum equals the net force", or $\vec{F} = m\vec{a}$ for a constant mass particle.

Observers that have a non-zero acceleration with respect to an inertial observer are called non-inertial. When the second law of motion is written in a non-inertial frame, the equation of motion is modified. The new equations of motion contain forces not present in the inertial frame are called fictitious or inertial forces. We will study these modifications in this chapter.

A particularly important application of these modified equations is to the frames fixed to the Earth, also called Earth-based frames. Since Earth is rotating about its axis, all observers on the Earth are accelerating with respect to any fixed inertial frame. This makes all Earth-based frames non-inertial.

Newton's second law of motion in rotating frames that have a constant angular velocity with respect to an inertial frame, two inertial forces arise. These inertial forces are called centrifugal and Coriolis forces. We will study their implications in some detail in this chapter. But, first we will study a simple situation of a frame accelerating in a straight line with respect to an inertial frame.

1.1 ACCELERATING FRAME

1.1.1 Kinematics in Accelerating Frame

The position of a particle in an accelerating frame is defined in the same way as in any other frame. The position vector of a point particle is the displacement vector from the origin to the current position of the particle. The relation with a non-accelerating frame can be established by drawing the two frames and noting the triangle of vectors formed by the position of the particle $\vec{r}$ and $\vec{r}\,'$ in the non-accelerating and accelerating frames respectively and the vector $\vec{R}$ from the origin O of the non-accelerating frame to the origin O' of the accelerating frame (Fig 1.1).

$$\vec{r} = \vec{R} + \vec{r}\,' \qquad (1.1)$$

The velocity $\vec{V}$ and acceleration $\vec{A}$ of the accelerating frame with respect to the non-accelerating frame are simply the velocity and acceleration of origin O' of the accelerating frame, and can be obtained

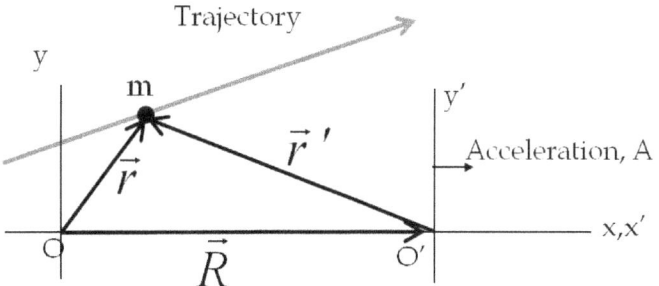

Fig. 1.1: Position vectors of a point mass in two frames.

by successively taking time derivatives of vector $\vec{R}$.

$$\vec{V} = \frac{d\vec{R}}{dt} \tag{1.2}$$

$$\vec{A} = \frac{d\vec{V}}{dt} = \frac{d^2\vec{R}}{dt^2} \tag{1.3}$$

The velocity $\vec{v}'$ of the particle with respect to the accelerating frame is simply the rate at which position $\vec{r}'$ of the particle changes with time.

$$\vec{v}' = \frac{d\vec{r}'}{dt} \tag{1.4}$$

Similarly, the acceleration $\vec{a}'$ of the particle with respect to the accelerating frame is simply the rate at which velocity $\vec{v}'$ of the particle with respect to the same frame changes with time.

$$\vec{a}' = \frac{d\vec{v}'}{dt} \tag{1.5}$$

Now, by taking successive derivatives of both sides of Eq. 1.1 with respect to time gives us the relation between velocity and acceleration of a particle in an in an inertial and a non-inertial frame.

$$\boxed{\vec{v} = \vec{V} + \vec{v}'} \tag{1.6}$$

$$\boxed{\vec{a} = \vec{A} + \vec{a}'} \tag{1.7}$$

1.1.2 Newton's Second Law in Accelerating Frame

Newton's second law of a point particle of mass m in an inertial frame is given by

$$m\vec{a} = \vec{F} \tag{1.8}$$

where $\vec{F}$ is the net force on the particle and $\vec{a}$ the acceleration. This equation remains the same in all inertial frames, which are frames that have constant velocity with respect to each other. Now, we wish to find the corresponding equation of motion if the particle's motion is observed from an accelerating frame. Let $\vec{a}\,'$ be the acceleration of the particle as observed from an accelerating frame. Let the acceleration of the accelerating frame with respect to the inertial frame be $\vec{A}$. By substituting Eq. 1.7 from the last section, we find that the following relation holds.

$$m\vec{a}\,' = \vec{F} - m\vec{A}. \tag{1.9}$$

Therefore, mass times acceleration of a particle with respect to an accelerating frame is not equal to the net force, but to the net force minus mass of the particle times acceleration of the frame itself. Thus, in an accelerating frame one needs to add $(-m\vec{A})$ to the real net force $\vec{F}$ to equate the "corrected" force to mass times acceleration.

The quantity $(-m\vec{A})$ acts like an additional force on the mass m, and is called a "**fictitious force**" or "**inertial force**" to distinguish it from the "real force" $\vec{F}$. In the accelerating frame the inertial forces are felt the same way as real forces except there is no agent that applies them. The inertial force in a uniformly accelerating frame acts just like the gravitational force since it is proportional to mass. This observation led Albert Einstein to develop an alternate theory of gravitation called the general theory of relativity.

Example 1.1. Apparent Weight in an Elevator. *As an example of an accelerating frame, consider observation made by a person in an elevator which accelerates with respect to the ground. We treat the frame of a ground-based observer as an inertial frame for purposes of this example. Both the ground-based observer and the observer in the elevator have access to a reading of a weighing scale fixed to the elevator. A person stands on the weighing scale in the elevator. What will be the reading in the weighing scale as recorded by the two observers?*

Solution. First, we note that the reading in the scale is that of the normal force between the scale and the person. Since the force is a real force, the two frames will have the same values for the normal force, although the calculations in the two frames will differ. In the following we present the two calculations.

Ground-based frame.

This is the non-accelerating frame, therefore there will be no inertial forces in this frame. The person on the scale has acceleration $\vec{A}$ with respect to this frame since the acceleration of the person is same as that of the elevator. The forces on the person are (1) gravity (Mg, pointed down) and (2) normal (unknown, pointed up). The free-body diagram is shown in Fig. 1.2.

Let y-axis be pointed up. Now, we write down the y-component of Newton's second law in this inertial frame.

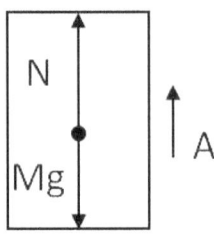

Fig. 1.2: The free-body diagram of forces on a person in an accelerating elevator drawn from the perspective of the ground-based frame. The person accelerates with the elevator.

$$N - Mg = MA$$

Therefore, the reading on the scale will be:

$$N = M(g + A).$$

This result says that, when the accelerator is accelerating up (meaning the direction of the acceleration is pointed up), the scale will give a higher reading than Mg. If the acceleration of the elevator is pointed down, then A will be negative. In that case the scale will read less than Mg. Note that the reading on the scale does not depend upon the direction of the motion of the elevator but rather the direction of the acceleration. Therefore the scale reading is more than Mg if the acceleration is pointed up regardless of whether the elevator is going up or going down.

Elevator frame.

This is non-inertial frame when the elevator has non-zero acceleration $\vec{A}$ with respect to the ground. In this frame the real forces on the person are the same two force, viz. the forces of gravity (Mg, down) and normal (N unknown, up). But the statement of the second law is different. We need to subtract $M\vec{A}$, from the real forces to get the net force on the person as shown in the free-body diagram of the person in the elevator frame given in Fig. 1.3.

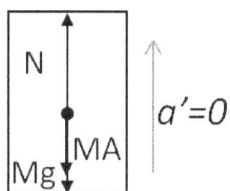

Fig. 1.3: The free-body diagram of forces on a person in an accelerating elevator drawn from the perspective of the elevator frame. Since the person accelerates with the elevator, his acceleration with respect to the elevator is zero.

The person on the scale does not have any acceleration with respect to this frame since the person moves with the elevator.

$$\vec{a}\,' = 0.$$

Taking the vertically up direction as the positive y-axis, the y-component of the modified equation of motion is

$$N - Mg - MA = 0$$

Therefore, we find $N = M(g + A)$, which is the same conclusion for the reading on the scale as we found by working in the ground-based inertial frame.

Example 1.2. Pendulum In An Accelerating Train.

As a second example, consider a pendulum of mass M hanging from the ceiling of a train. When the train is at rest or coasting at a constant velocity, the pendulum hangs vertically. But when the train is accelerating the bob hangs at an angle θ to the vertical. We wish to find this angle when the train's acceleration is $\vec{A}$. We will do this problem in two frames to illustrate a non-accelerating frame and an accelerating frame.

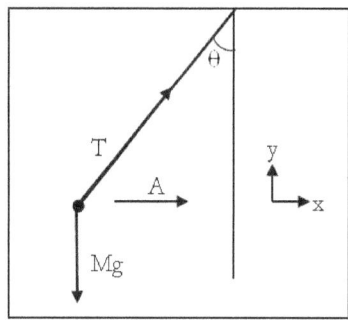 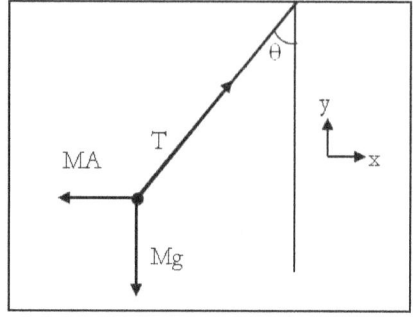

(a) Inertial Ground Frame   (b) Accelerating Train Frame

Fig. 1.4: Example 1.2. Free-body diagrams in (a) inertial and (b) non-inertial frames. Note the additional inertial force on the bob in the accelerating frame.

Equations of motion in the ground-based frame ($\vec{a} \neq 0$):

$$\text{x component: } T\sin\theta = MA$$
$$\text{y component: } T\cos\theta - Mg = 0$$

Equations of motion in the train-based frame ($\vec{a}\,' = 0$):

$$\text{x component: } T\sin\theta - MA = 0$$
$$\text{y component: } T\cos\theta - Mg = 0$$

Note that the two frames yield identical relations for the equations of motion. Both frames will yield the same value for the real forces such as the tension in the string.

1.2 ROTATING FRAME

1.2.1 Kinematics in a Uniformly Rotating Frame

Since a rotating frame is not an inertial frame, the form of the second law, $\vec{F} = m\vec{a}$, is modified to $\vec{F} - m\vec{A} = m\vec{a}\,'$, where $\vec{A}$ is the acceleration of the non-inertial frame with respect to the inertial frame. In this section, we will work out the implications of the acceleration of the frame arising from the rotation of the frame with respect to an inertial frame.

Let us denote the kinematic variables in the accelerating frame with a prime as above. Let $\vec{r}\,'$, $\vec{v}\,'$, and $\vec{a}\,'$ be the position, velocity, and acceleration of the particle as observed in the rotating frame.

Their definitions are the same in the rotating frame as in any other frame, viz., the velocity being the rate of change of the position and the acceleration the rate of change of the velocity.

To relate the position, velocity and acceleration vectors in the rotating frame to the corresponding quantities in the non-rotating inertial frame, it is helpful to examine these quantities for a point particle in two different frames that share a common origin but one rotating with respect to the other about the common z-axis as shown in Fig. 1.5.

For simplicity we pick the zero of time at the instant when the two frames have their axes lined up in the same directions so that all the coordinates of the same particle in the two frames have the same values at $t = 0$. In the following, we will consider two situations. In the first situation, we will choose a particle at rest in the rotating frame, and in the second situation, the particle will be allowed to move in a plane perpendicular to the rotation axis of the rotating frame.

A particle fixed at the x-axis of a uniformly rotating frame

Consider a particle fixed at $(x', 0, 0)$ as shown in Fig. 1.5. Although this particle's position in the rotating frame is fixed, its position in the fixed frame is changing. As a matter of fact, as far as the fixed frame is concerned, the particle moving in xy-plane in a circle of radius equal to x' with angular speed Ω.

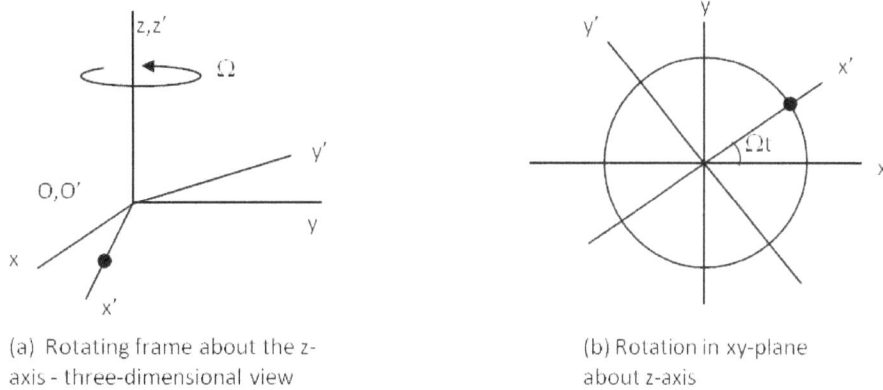

(a) Rotating frame about the z-axis - three-dimensional view

(b) Rotation in xy-plane about z-axis

Fig. 1.5: Looking at a particle at rest in a rotating frame.

The position of the particle in the two frames can be obtained by simple trigonometry of the situation in the xy-plane. We will omit the z-components since they are all zero in the two frames.

$$\text{Rotating frame: } x' = \text{fixed}, \ y' = 0$$
$$\text{Inertial frame: } \quad x = x' \cos \Omega t, \ y = x' \sin \Omega t$$

Therefore, the velocities of the particle in the two frames, obtained by taking derivatives of the positions, are as follows.

$$\text{Rotating frame: } \quad v'_x = \frac{dx'}{dt} = 0, \ v'_y = \frac{dy'}{dt} = 0$$
$$\text{Inertial frame: } \quad v_x = \frac{dx}{dt} = -\Omega x' \sin \Omega t, \ v_y = \frac{dy}{dt} = \Omega x' \cos \Omega t$$

The acceleration of the particle in the two frames can be obtained by taking derivatives of the corresponding velocities.

$$\text{Rotating frame: } \quad a'_x = \frac{dv'_x}{dt} = 0, \quad a'_y = \frac{dv'_y}{dt} = 0$$
$$\text{Inertial frame: } \quad a_x = \frac{dv_x}{dt} = -\Omega^2 x' \cos \Omega t, \quad a_y = \frac{dv_y}{dt} = -\Omega^2 x' \sin \Omega t$$

The x and y-components of acceleration in the inertial frame are equivalent to the radially inward acceleration, i.e. the centripetal acceleration.

Inertial frame:

$$\vec{a} = \begin{cases} \text{Magnitude} = \Omega^2 x' \\ \text{Direction: Towards the origin.} \end{cases} \tag{1.10}$$

This makes sense, because from the perspective of the fixed frame, the particle moves in a uniform circular motion of radius equal to x' at constant angular speed Ω. Therefore, the acceleration in the fixed frame must be pointed radially inwards.

Vector Notation

It is instructive to write the kinematic quantities in the fixed and rotating frames given above in a vector notation. The example particle does not move with respect to the rotating frame. This particle will appear to rotate in a circle about the z-axis when observed from the fixed frame. Therefore, in the fixed frame the angular velocity vector will be pointed along the z-axis.

$$\vec{\Omega} = \Omega \hat{u}_z. \tag{1.11}$$

The velocity of the particle in the fixed frame is tangential to the circle in the xy-plane and is given by the cross product of the angular velocity vector and the position vector.

$$\vec{v} = \frac{d\vec{r}}{dt} = \vec{\Omega} \times \vec{r}. \tag{1.12}$$

Since acceleration is equal to the time-derivative of velocity, we obtain the following for the acceleration in the fixed frame.

$$\vec{a} = \frac{d\vec{v}}{dt} == \vec{\Omega} \times \frac{d\vec{r}}{dt}$$
$$\implies \vec{a} = \vec{\Omega} \times \left(\vec{\Omega} \times \vec{r} \right) \quad (\text{Const } \vec{\Omega})$$

We find that, although the particle is not moving in the rotating frame, the particle has an acceleration in the fixed frame.

A particle moving in the xy-plane of a uniformly rotating frame

Let us consider a particle that is not fixed but moves in the xy-plane of a uniformly rotating frame $Ox'y'z'$ which is rotating with respect to a fixed frame $Oxyz$ about their common z-axis. Since the frame is rotating about the z-axis, the particles moves in the xy-plane of the fixed frame as well. The situation between time t and $t + \Delta t$ is shown in Fig. 1.6. For simplicity, let the axes of the two frames be coincident at time t.

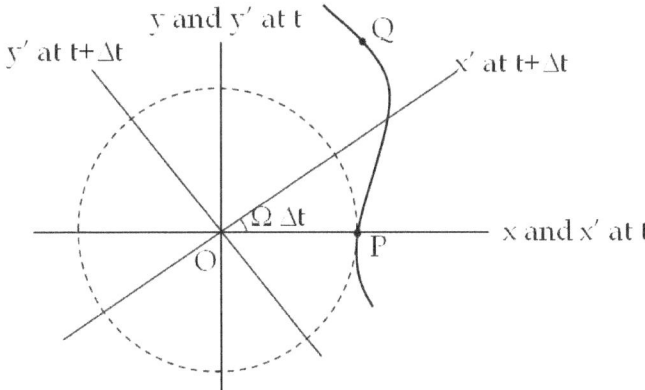

Fig. 1.6: The physical situations at times t and $t + \Delta t$. In duration t to $t + \Delta t$, the particle moves from P to Q, and the rotating frame rotates by an angle Ωt. The common z-axis pointed out-of-page.

Let (x, y) and (x', y') be the coordinates of the particle at some time t in the $Oxyz$ and $Ox'y'z'$ frames respectively. The change of the coordinates over the interval from t to $t + \Delta t$ are $(\Delta x, \Delta y)$ and $(\Delta x', \Delta y')$ respectively. What are their relations? The relation between the displacements in the two frames is more conveniently written in the vector notation. Therefore, we will work with vector notation here.

At time t, let the particle be at point P along momentarily coincident x and x'-axes (Fig. 1.7). The non-rotating frame marks the

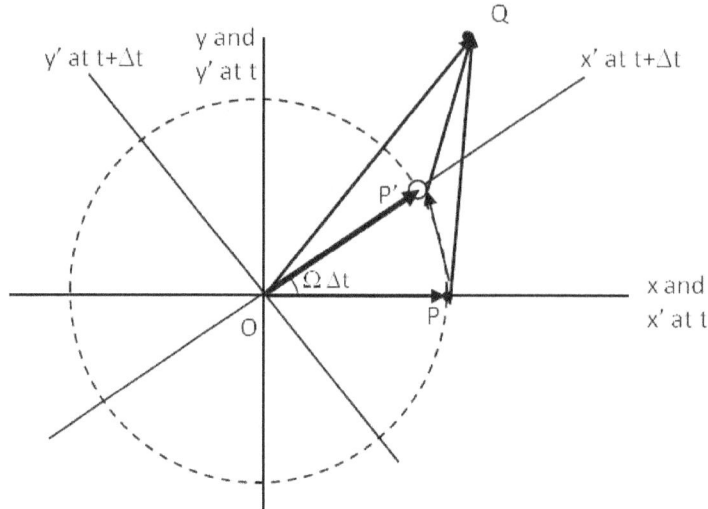

Fig. 1.7: Position of a point particle P from rotating frame.

location with a marker at P at its x-axis and the rotating frame marks the same location in its frame as P' at its x-axis. As time passes, the particle moves to another location shown as Q at time $t + \Delta t$. Meanwhile the rotating frame has also moved.

Therefore, from the perspective of the rotating frame, the displacement of the particle between t and $t + \Delta t$ is the vector from P' on its x-axis where the particle was at t to the space point Q. The same displacement of the particle from the perspective of the non-rotating frame is PQ vector in space. From $\triangle PP'Q$ we obtain the following relation in space.

$$\overrightarrow{PQ} = \overrightarrow{PP'} + \overrightarrow{P'Q}. \tag{1.13}$$

Now, $\overrightarrow{PP'}$ is on the arc of a circle of radius $|\vec{r}|$ with the arc angle $\Omega \Delta t$, therefore

$$\overrightarrow{PP'} = \left(\vec{\Omega} \Delta t \right) \times \vec{r}.$$

The vector $\overrightarrow{PQ}$ is the displacement $\Delta \vec{r}$ of the particle in the fixed frame and $\overrightarrow{P'Q}$ is the displacement $\Delta \vec{r}\,'$ of the particle in the rotating frame. Putting these quantities in Eq. 1.13 we have the following relation among displacements in the time interval Δt.

$$\Delta \vec{r} = \left(\vec{\Omega} \Delta t \right) \times \vec{r} + \Delta \vec{r}\,'.$$

Dividing by Δt and taking the $\Delta t \to 0$ limit, we find that the velocity in the inertial frame is equal to the sum of the velocity of the particle in the rotating frame and an additional term resulting from the rotation of the frame.

$$\vec{v} = \vec{v}\,' + \vec{\Omega} \times \vec{r}. \tag{1.14}$$

Rather than use prime for the quantities in the rotation frame, sometimes a different notation is use in which we attach a subscript "in" and "rot" for quantities in the fixed frame and the rotating frame respectively.

$$\vec{v}_{in} = \vec{v}_{rot} + \vec{\Omega} \times \vec{r}. \tag{1.15}$$

There is no need to put a subscript to $\vec{r}$ since $\vec{r}$ is a vector from the origin to the position of the particle and the origins of the two frames are at the same point. The result of relation between the velocity vectors is also often written as

$$\left(\frac{d\vec{r}}{dt}\right)_{in} = \left(\frac{d\vec{r}}{dt}\right)_{rot} + \vec{\Omega} \times \vec{r}. \tag{1.16}$$

where the subscript "in" denotes the quantity in the inertial or non-rotating frame and the subscript "rot" for a quantity with respect to the rotating frame.

Note that to derive this relation, we only used the geometric property of the position vector. A similar argument can be made about the rate of change of any arbitrary vector $\vec{w}$ in the two frames.

$$\boxed{\left(\frac{d\vec{w}}{dt}\right)_{in} = \left(\frac{d\vec{w}}{dt}\right)_{rot} + \vec{\Omega} \times \vec{w}.} \tag{1.17}$$

For instance, we can obtain the time-derivative of the velocity vector $\vec{v}_{in}$ in the inertial frame by setting $\vec{w} = \vec{v}_{in}$ in this equation.

$$\boxed{\left(\frac{d\vec{v}_{in}}{dt}\right)_{in} = \left(\frac{d\vec{v}_{in}}{dt}\right)_{rot} + \vec{\Omega} \times \vec{v}_{in}.} \tag{1.18}$$

The left hand side of this equation is the acceleration of the particle in the inertial frame since it is the rate of change of the corresponding velocity in the same frame. The first term on the right side of this equation is not acceleration of anything since this term mixes information from the two frames - this term is the rate of change of the velocity in the inertial frame as observed from the rotating frame.

Finally, by substituting $\vec{v}_{in}$ in the right-side of Eq.rot-inert-eq-2 in terms of $\vec{v}_{rot}$ as given in Eq. 1.15 we can work out the relation between accelerations in the two frames.

$$\left(\frac{d\vec{v}_{in}}{dt}\right)_{in} = \left[\frac{d}{dt}\left(v_{rot} + \vec{\Omega} \times \vec{r}_{in}\right)\right]_{rot} + \vec{\Omega} \times \left(\vec{v}_{rot} \times \vec{\Omega} \times \vec{r}_{in}\right)$$

$$= \vec{a}_{rot} + 2\vec{\Omega} \times \vec{v}_{rot} + \vec{\Omega} \times \left(\vec{\Omega} \times \vec{r}\right)$$

Therefore, the accelerations of the particle in the two frames are related as follows.

$$\boxed{\vec{a}_{in} = \vec{a}_{rot} + 2\vec{\Omega} \times \vec{v}_{rot} + \vec{\Omega} \times \left(\vec{\Omega} \times \vec{r} \right)} \quad \text{(Constant } \vec{\Omega}). \qquad (1.19)$$

1.2.2 Newton's Second Law in Uniformly Rotating Frame

Newton's second law for a particle of fixed mass m is written in inertial frame as $m\vec{a}_{in} = \vec{F}$, where $\vec{F}$ is the net real force. Substituting $\vec{a}_{in}$ from Eq. 1.19 into the second law we find that mass times acceleration in a rotating frame has the following form.

$$\boxed{m\vec{a}_{rot} = \vec{F} - m \left[2\vec{\Omega} \times \vec{v}_{rot} + \vec{\Omega} \times \left(\vec{\Omega} \times \vec{r} \right) \right]} \quad \text{(Constant } \vec{\Omega})$$
$$(1.20)$$

Therefore, in a rotating frame, mass times acceleration is not equal to the net real force on the particle. In addition to the net real force $\vec{F}$, there are terms that have units of force but are dependent on the rotation of the frame. These are the "**fictitious forces**" or "**inertial forces**". In a rotating frame, the "real" and "fictitious" forces are indistinguishable in their effect on the acceleration except that there appears to be no agent(s) for the fictitious forces as far as the rotating frame is concerned.

The first term $\left(-2m\vec{\Omega} \times \vec{v}_{rot} \right)$ is called the **Coriolis force**, and the second term $\left[-m\vec{\Omega} \times \left(\vec{\Omega} \times \vec{r} \right) \right]$ the **centrifugal force**. While the direction of the Coriolis force is perpendicular to the axis of rotation and the velocity of the particle, the centrifugal force is always pointed away from the axis of rotation and remains perpendicular to the axis.

1.2.3 Newton's Second Law in Earth's Frame

An Earth-based frame is a rotating frame. Therefore, the equation of motion of a particle in an Earth-based frame will have Coriolis force and centrifugal force. The centrifugal force is included in mg, therefore, the equation of motion in the rotating frame of the Earth has Coriolis force only.

$$\boxed{m\vec{a}_{rot} = m\vec{g} + \vec{F}_{\text{other}} - 2m\vec{\Omega} \times \vec{v}_{rot},} \qquad (1.21)$$

where $\vec{F}_{\text{other}}$ are forces other than gravity and $\vec{g}$ is the acceleration due to gravity written as a vector to include the direction information

in the equation. We will use Eq. 1.21 to study the motion the motion
of particles near the surface of the Earth in an Earth-based frame.

Example 1.3. Freely Falling Particle In An Earth-Based Frame.
*The rotation of earth has observable effects on a freely falling object.
If a particle of mass m is released at rest from a height h above the
surface of the Earth, it will fall towards the center of the Earth if the
Earth were not rotating. But, because the Earth is rotating, the path of
the particle will deviate from this direction in an Earth-based frame.
In this example we wish to determine the deviation at the equator.*

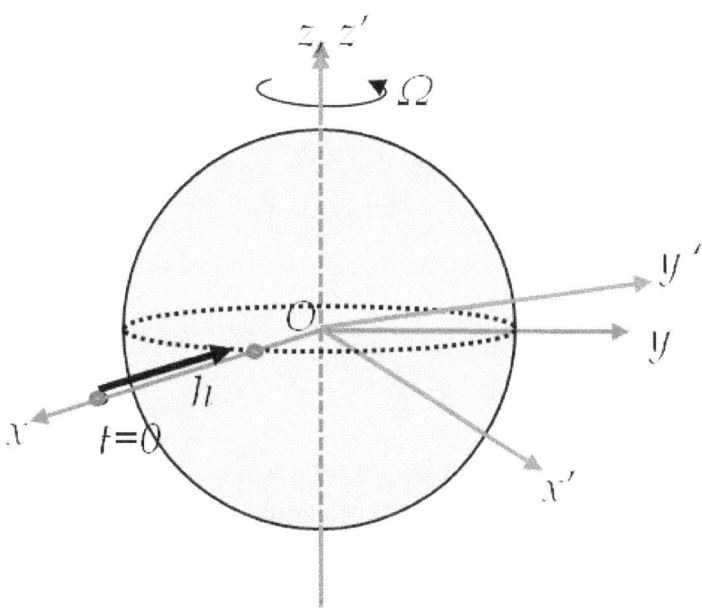

Fig. 1.8: Example 1.3. The rotating frame $O'x'y'z'$ and the fixed frame $Oxyz$
coincide at $t = 0$. That is, Ox' and Ox are in the same direction at $t = 0$ when
the particle is released on the x-axis. The particle falls down the x-axis of the
fixed coordinate system.

Solution. Consider two coordinate frames $Oxyz$ and $Ox'y'z'$ with
their origins at the center of Earth that have the z-axis pointed in
the direction of the axis of rotation of the Earth. Let the frame $Ox'y'z'$
rotate with respect to the fixed inertial frame $Oxyz$ at angular speed
Ω, and let the axes of the two systems be coincident at $t = 0$. We will
use M for the mass of the Earth and assume Earth to be a sphere of
radius R.

Suppose now a particle of mass m is released at $t = 0$ from rest at
$x = x' = h + R$. In the inertial frame the particle moves straight down
the x-axis and reaches $x = R$ at some time Δt. During Δt the x' and
y' axes of the rotating frame move out to another direction. Since the
rotation is about the z-axis, the particle will fall in the $x'y'$ plane of
the rotating frame. We wish to determine the y' component of the

displacement when the particle has dropped a distance of h in the inertial frame. We start by writing the components of the equation in the rotating frame.

$$\frac{dv'_x}{dt} = -g - \left(2\vec{\Omega} \times \vec{v}_{rot}\right)_{x'} \tag{1.22}$$

$$\frac{dv'_y}{dt} = 0 - \left(2\vec{\Omega} \times \vec{v}_{rot}\right)_{y'} \tag{1.23}$$

$$\frac{dv'_z}{dt} = 0 - \left(2\vec{\Omega} \times \vec{v}_{rot}\right)_{z'} \tag{1.24}$$

The x' and y' components of the Coriolis and centrifugal terms are as follows.

$$\vec{\Omega} \times \vec{v}_{rot} = \begin{vmatrix} \hat{u}_{x'} & \hat{u}_{y'} & \hat{u}_{z'} \\ 0 & 0 & \Omega \\ v_{x'} & v_{y'} & v_{z'} \end{vmatrix} = -\Omega v_{x'} \hat{u}_{y'} + \Omega v_{x'} \hat{u}_{z'}.$$

Hence the equations of motion of the particle are:

$$\frac{dv_{x'}}{dt} = -g - 2\Omega v_{y'} \tag{1.25}$$

$$\frac{dv_{y'}}{dt} = 2\Omega v_{x'} \tag{1.26}$$

$$\frac{dv_{z'}}{dt} = 0 \tag{1.27}$$

Solving these equations with the initial condition $x' = h + R$, $y' = 0$, $z' = 0$, and $v_{0x'} = v_{0y'} = v_{0z'} = 0$ will give us the trajectory of the particle from the perspective of the rotating frame, i.e., from someone observing the particle from Earth-based frame. Since there is no motion along the z-axis, we will work out the solution for x' and y' components only.

Approximate solution: Observe that the particle will pick up velocity more along the vertical direction than along the horizontal direction. Therefore, we can assert that in time t, $|v'_x| \approx gt$. Using this in v'_x equation, we find the following for v'_y.

$$\frac{dv_{y'}}{dt} \approx -2\Omega gt \implies v_{y'} = -\Omega gt^2,$$

where we have used the initial condition on $v_{y'}$. Integrating $v_{x'}$ and $v_{y'}$ we obtain the following for x' and y' coordinates after the particle is released at ($x' = h + R$, $y' = 0$) at $t = 0$.

$$x' = h + R - \frac{1}{2}gt^2 \tag{1.28}$$

$$y' = -\frac{1}{3}gt^3 \tag{1.29}$$

From the x' equation we can determine the time T for the particle to the surface of Earth, which has $x' = R$. Using this time in the y' equation we find the horizontal deviation y'.

$$\Delta y' = -\frac{1}{3}\Omega g \left(\frac{2h}{g}\right)^{3/2}$$

For a 100 meter drop we will find the deviation to be

$$\Delta y' = -\frac{1}{3}\frac{2\pi}{24 \times 3600\ s} \times 9.81\ \text{m/s}^2 \times \left(\frac{2 \times 100\ \text{m}}{9.81\ \text{m/s}^2}\right)^{3/2}$$
$$= -2.2 \times 10^{-2}\ \text{m}.$$

Since the rotation axis is towards the North, the y' axis is towards the East at the surface of the Earth. Therefore, a particle dropped from 100 m above the ground will land approximately 2.2 cm to the West of the line connecting the original position to the center of the Earth.

Fig. 1.9: Foucault pendulum. The rotation of the Earth causes the plane of oscillation of the pendulum to change over time. With changing plane of oscillation, the pins at different positions are knocked down at different times. Photo credit: Ciudad de las Artes y de las Ciencias de Valencia by Daniel Sancho, Wikicommons.

Example 1.4. Surface of a Rotating Fluid in a Bucket.

A bucket of water is rotated with a uniform rotational speed Ω. It is found the surface assumes a steady shape. Determine the shape. Ignore the rotation of earth.

Solution. Due to the symmetry in the problem, it is sufficient to work in a plane containing the axis of rotation and a horizontal direction as shown in Fig. 1.10. We will call the axis of rotation the

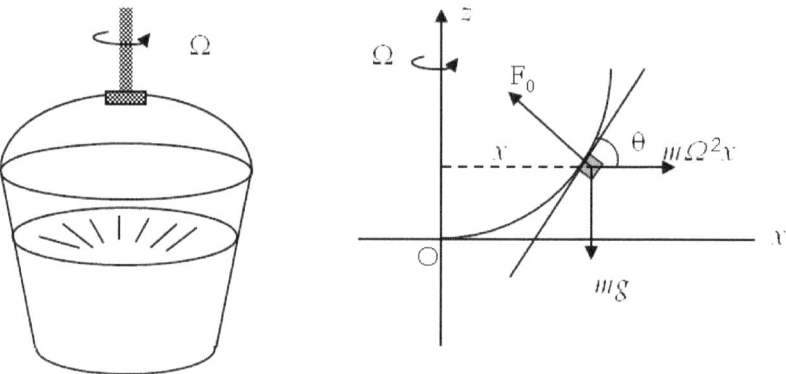

Fig. 1.10: Example 1.4.

z-axis and the horizontal direction will be taken to be the x-axis. Therefore, to find the equation of the surface, we need to work out the function $z(x)$. We will make use of steady condition on a mass element at the surface.

Here, notice that it is easier to work in the rotating frame of the bucket since in this frame, liquid in the bucket will not be moving and will have zero acceleration. That is, in this frame, real forces will be balanced by inertial force(s). Let us figure out the real and inertial forces on a mass element at the surface.

In the rotating frame once the steady state has reached, water would not be moving any more. Therefore the Coriolis force will be zero. Thus, the only inertial force on particles of water will be the centrifugal force. The only other force is the weight.

Consider a mass element of mass m at the surface of the steady fluid. The weight has magnitude mg and acts straight down. The centrifugal forces on various water particles and the weight of the water molecules will press on the layers of water in contact. The water molecules on the surface will press on the molecules just below the surface. The reaction force from the layer just below the surface will be in the normal direction of the surface as shown by F_0 in the figure.

Therefore, the x and z-components of the equations of motion of an element at the surface of water in the rotating frame for a mass at the surface are:

$$x\text{-component:} \quad m\Omega^2 x - F_0 \sin\theta = 0$$

$$z\text{-component:} \quad F_0 \cos\theta - mg = 0$$

Therefore,

$$\tan\theta = \frac{\Omega^2}{g} x$$

But this tangent must equal the slope of the tangent to the curve $z(x)$ at the point.

$$\frac{dz}{dx} = \tan\theta.$$

Therefore, we obtain the following equation for $z(x)$.

$$\frac{dz}{dx} = \frac{\Omega^2}{g} x,$$

which can be immediately integrated to yield

$$z = \frac{\Omega^2}{2g} x^2 + C,$$

where C is the constant of integration. From the figure, $x = 0$ corresponds to $z = 0$ on the surface, therefore $C = 0$. Hence, the equation for the surface is

$$z = \frac{\Omega^2}{2g} x^2.$$

The situation is symmetric in the xy-plane and there is nothing special about x-axis. To obtain the equation for the entire surface and not just a slice of the surface, all we need to do is to replace x by the radial distance r of polar coordinates.

$$z = \frac{\Omega^2}{2g} r^2. \tag{1.30}$$

Hence, the surface is a paraboloid of revolution about the axis of rotation.

1.3 CORIOLIS FORCE

In the Earth-based frame, there are two inertial forces, one directed away from the center of earth, and the other pointed in the plane perpendicular to the axis of rotation. The centrifugal force subtracts from the centrally directed force of gravitation, and is absorbed in the value for the acceleration due to gravity g. Therefore, the inertial force that leads to unexpected effects in the Earth-based frame is the Coriolis force $\vec{F}_{cor}$.

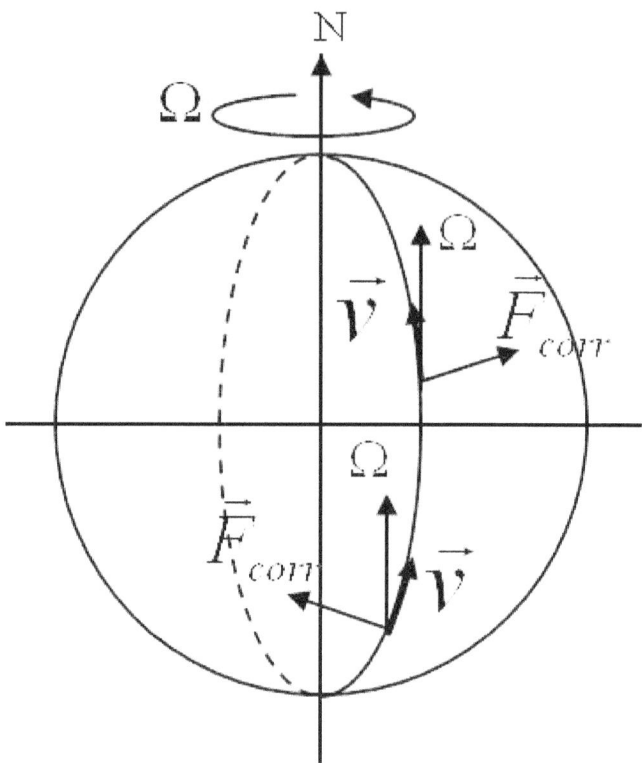

Fig. 1.11: Coriolis force in northern and southern hemisphere.

$$\boxed{\vec{F}_{cor} = -2m\vec{\Omega} \times \vec{v}.} \qquad (1.31)$$

Note that the Coriolis force is perpendicular to the direction of velocity of the particle, as evident from the cross product. Therefore, a moving particle is deflected perpendicular to its direction of motion by the Coriolis force. As a result, northward moving particles in the northern hemisphere are deflected to the East while northward moving particles in the southern hemisphere are deflected to the West as illustrated in Fig. 1.11.

The atmosphere of Earth is usually modeled as particles in motion. The forces on particles of atmosphere are from gravity and pressure differences. If there is a pressure gradient, then there is a force from high pressure region towards the low pressure region. Due to the Coriolis force, when a particle has nonzero velocity towards the lower pressure region, the path of the particle is changed towards the perpendicular direction. As a result the winds circulates differently in the northern and souther hemispheres, being in the counterclockwise sense in the northern hemisphere and in the clockwise sense in the southern hemisphere as shown in Figs. 1.12 and 1.13 respectively.

Fig. 1.12: Hurricane Katrina in the Gulf of Mexico rotating counterclockwise on August 28, 2005. Hurricane Katrina later devastated the city of New Orleans in the state of Louisiana causing massive flooding. Credit: National Oceanic and Atmospheric Administration, USA.

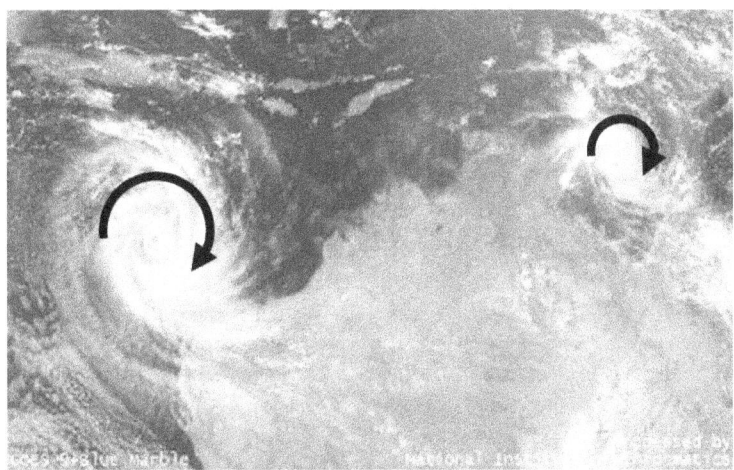

Fig. 1.13: Rotation of winds in southern hemisphere. Cyclones Willy (left) and Ingrid (right) off the northern coast of Australia on 11th March 2005. (Credits: National Institute of Informatics, Japan).

1.4 EXERCISES

Accelerating Frame

Ex 1.1. Two friends John and Jane observe the location of a building from their own frames which are accelerating with respect to each other. In Jane's frame John has a constant acceleration $\vec{A}$ whose direction makes an angle θ with respect to the direction of the building. Let the building be at a distance R from Jane who is fixed to the Earth.

Suppose at $t = 0$, John and Jane were at the same location and their relative velocity was zero. (a) Draw a figure in Jane's frame

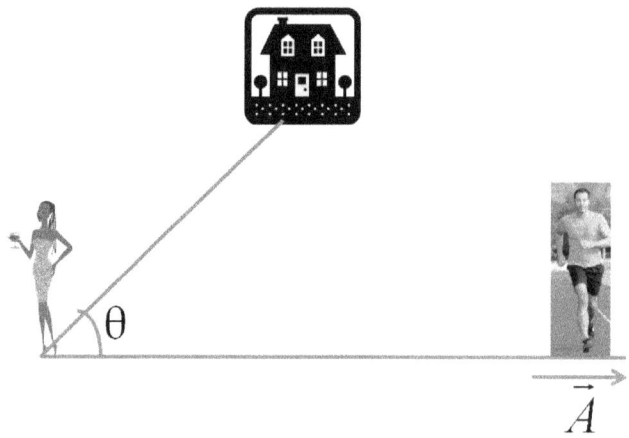

Fig. 1.14: Exercise 1.1.

showing a choice of coordinates for both John and Jane supposing that the position vector of the building and acceleration vector $\vec{A}$ are in the xy-plane. You can also take the x-axis to be the direction of vector $\vec{A}$. (b) Deduce the relation between the coordinates of the building in the two frame at an arbitrary time? (c) Describe the motion of the building in the two frames.

Ex 1.2. A box sits on a rubber floor in a truck. The driver starts the truck and increases the acceleration steadily from zero to 5 m/s^2 in 2 seconds. If the coefficient of static friction between the box and the rubber floor is 0.2, determine if the box will slide, and if so, at what acceleration? Do this problem in the frame of the truck.

Ex 1.3. The coordinates of a particle in space from two frames $Oxyz$ and $O'x'y'z'$ are related as follows.

$$x' = x + (5 \text{ m/s}) \, t$$
$$y' = x + (5 \text{ m/s}) \, t + (2 \text{ m/s}^2) \, t^2$$
$$z' = z$$

(a) Describe the relative motion of the two frames with respect to each other. (b) If a particle has a constant velocity with respect to $Oxyz$ with components $(2 \text{ m/s}, 0, 0)$, what are the velocity and acceleration of this particle with respect to $O'x'y'z'$? (c) If a particle has a constant velocity with respect to $O'x'y'z'$ with components $(2 \text{ m/s}, 0, 0)$, what are the velocity and acceleration of this particle with respect to $Oxyz$?

Ex 1.4. A box is sliding on a floor so that its coordinates with respect to a frame $Oxyz$ fixed with respect to the floor are given as

$$x = (1 \text{ m}) + (2 \text{ m/s}) \, t + (3 \text{ m/s}^2) \, t^2$$
$$y = z = 0$$

What would be the coordinates of this box when observed with respect to another frame $O'x'y'z'$ that has the following acceleration with respect to $Oxyz$?

$$A_x = 2 \text{ m/s}^2$$
$$A_y = A_z = 0$$

Assume origins O and O' coincide at $t = 0$ and the axes of the two frames are parallel with each other.

Ex 1.5. A box is sliding on a floor so that its coordinates with respect to a frame $Oxyz$ fixed with respect to the floor are given as

$$x = (1 \text{ m}) + (2 \text{ m/s}) \, t + (3 \text{ m/s}^2) \, t^2$$
$$y = 0$$
$$z = 0$$

What would be the coordinates of this box when observed with respect to another frame $O'x'y'z'$ that has the following acceleration with respect to $Oxyz$?

$$A_x = 0$$
$$y = 2 \text{ m/s}^2$$
$$z = 0$$

Assume the origins O and O' coincide at $t = 0$ and the axes of the two frames are parallel with each other.

Ex 1.6. A block of mass m is hung from the ceiling of an elevator using a spring. Find the change in the length of the spring under the following situations. (a) Elevator going up with constant velocity, $\vec{v}_1$. (b) Elevator going down with constant velocity, $\vec{v}_2$. (c) Elevator going up with acceleration $\vec{a}_1$ point up. (d) elevator going up with acceleration $\vec{a}_2$ pointed down. (e) elevator going down with acceleration $\vec{a}_3$ pointed up. (f) elevator going down with acceleration $\vec{a}_4$ pointed down.

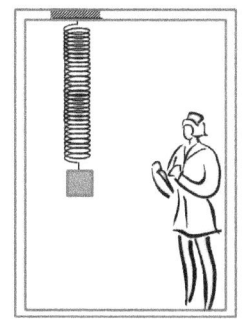

Fig. 1.15: Exercise 1.6.

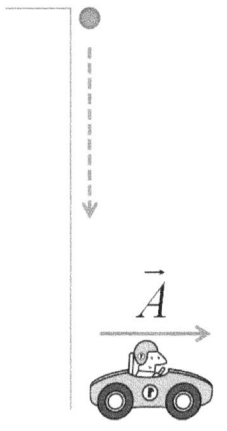

Fig. 1.16: Exercise 1.7.

Ex 1.7. A ball is dropped from a 50-m building. In a frame fixed to Earth, the ball drops vertically with a constant acceleration g pointed down. The same ball is observed from a car that is accelerating with respect to the fixed frame. The acceleration of the car is pointed horizontally and away from the building.

Suppose the fixed frame has a coordinate system $Oxyz$ whose y-axis is pointed up and the x-axis is pointed in the direction of the acceleration of the car. Let $O'x'y'z'$ be the coordinate system whose origin is fixed to the car. The coordinate axes of the two frame are parallel to each other. The time $t = 0$ is chosen when the two origins coincide and the velocity of the car is zero with respect to the ground. (a) Write the coordinates of the ball in $Oxyz$ frame during the flight, i.e., before the ball hits the ground. (b) Write the coordinates of the ball in $O'x'y'z'$ frame during the flight, i.e., before the ball hits the ground. (c) What are the trajectory of the ball in the two frames? Do the trajectory equations make sense, i.e. does the ball fall vertically in both frames?

Fig. 1.17: Exercise 1.8.

Ex 1.8. A freely falling sky diver looks at another diver directly above him that has his parachute on and finds his position in a coordinate system that has the y-axis vertically up as follows.

$$y = 4 + 5t + 3t^2.$$

If at $t = 0$, the freely falling person was 100 meters from the ground. Find the position of the person with his parachute on as a function of time when observed from the ground?

Rotating Frame

Ex 1.9. A large platform is rotating uniformly with an angular speed Ω about the z-axis of a fixed frame $Oxyz$ with the origin at the center of the platform. A man is on the platform at a distance R from the center. (a) What is the motion of the man in the fixed frame? (b) What is the motion of the man with respect to a frame $O'x'y'z'$ that has the same origin and z-axis as the fixed frame but whose x and y-axes rotate with the platform? (c) Let the man be along the x' axis at $t = 0$ when he starts to walk directly towards the center at a constant speed v'. That is, he is walking along the x' axis towards origin with speed v' with respect to the origin O'. What is the motion of the person as seen from the fixed frame?

Ex 1.10. A car in Los Angeles, California, USA, is moving with a velocity of 30 m/s towards North with respect to an observer on the ground. Find the magnitude and direction of the Coriolis force on the car. Los Angeles, USA has latitude 34.0522° N, and longitude 118.2428° W.

Ex 1.11. A car in Canberra, Australia, is moving with a velocity of 30 m/s towards North with respect to an observer on the ground. Find the magnitude and direction of the Coriolis force on the car. Canberra, Australia has latitude 35.2828° S, and longitude 149.1314° E.

Ex 1.12. In exercise, Ex 1.9, write the equation of motion of the CM of the man with respect to (a) the fixed frame and (b) the rotating frame.

Ex 1.13. A satellite is in the Geocentric orbit about Earth at a distance R from the center of the Earth. The satellite revolves in a circular motion with the angular speed Ω, which is equal to the angular speed of the Earth about the same axis, as observed from a fixed frame. (a) Describe the motion of the satellite with respect to a frame with the origin at the center of Earth and which rotates with the Earth. (b) Write the equation of motion of the satellite in the two frames.

Ex 1.14. A bead of mass m can slide frictionlessly on a rotating ring of mass M and radius R which rotates at an angular speed of ω about a vertical axis through its center. When the bead is at a particular location, it does not slide. (a) Find this special position of the bead using calculations in a rotating frame. (b) Repeat the calculation in an inertial frame.

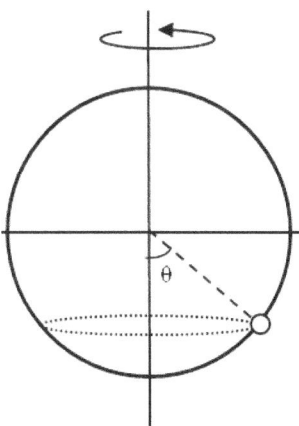

Fig. 1.18: Exercise 1.14.

Ex 1.15. A block of mass M is attached to a string of length l. The other end of the string is attached to a post at the center of a rotating platform. When the platform is rotating at a steady angular speed Ω the block moves in a circular motion. Find the radius of the circular motion of the block performing the calculations in a rotating frame.

Ex 1.16. A stone is dropped from the roof of a 200-m tall building at a place whose latitude is 30° N. Where will the stone land on the ground? (Ignore the effects of the air resistance.)

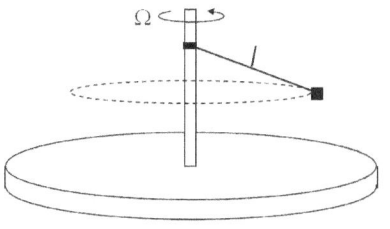

Fig. 1.19: Exercise 1.15.

1.5 PROBLEMS

P 1.1. A rocket of mass M is moving at constant velocity $\vec{v}$ in zero gravity environment. It ejects fuel from the back at a steady rate of $(dm/dt = \alpha)$ at speed u with respect to the rocket. What is the increase in speed of the rocket when a mass m_f of fuel has been ejected? Do this problem in the frame of the rocket.

P 1.2. A stone is tied to a string and rotated in a vertical circle at constant speed v inside an accelerating train. The circle of rotation is perpendicular to the train's velocity as observed from a frame outside the train. The train has a constant acceleration of magnitude A and direction East with respect to an observer outside the train. The train was at rest at $t = 0$. Find the tension in the string at four instances in the vertical circle of the motion of the stone: (a) when the stone is at the top, (b) when the stone is at the bottom, (c) when the stone is horizontal left, and (d) when the stone is horizontal right.

P 1.3. Calculate the percentage mistake made in evaluating the acceleration due to gravity when ignoring the rotation of earth?

P 1.4. Calculate the percentage mistake made in evaluating acceleration due to gravity when ignoring the revolution of earth around the sun?

P 1.5. You are inside a large enclosed container and everything in the entire container, including you, is rotating at a constant rate about some fixed axis, but you do not know the rate of rotation. To find the rate of rotation you place a penny at different places on a frictionless floor, and discover that the penny accelerates everywhere you place the penny. The directions of the acceleration of the penny at various place on the flat floor meet at a point X. (a) If the penny accelerates at 30 m/s^2 at a distance of 10 meters from the point X, what is the rate of rotation of the entire container? (b) What is the acceleration of the penny when it is at a point 20 meters from the point X?

P 1.6. A pendulum of length L and mass m in an accelerating frame is in equilibrium at an angle θ_0. (a) What is the acceleration of the accelerating frame with respect to an inertial frame? (b) When the pendulum is disturbed from the equilibrium angle by a small angle θ about θ_0, the pendulum oscillates. Determine the frequency of small oscillations in the accelerating frame.

P 1.7. A stone is dropped from rest from a height of 200 meters in a place at a latitude of 60 degrees and longitude 30 degrees. Where will it land on the surface of the Earth?

P 1.8. A stone is dropped from rest from a height of 200 meters in a place at a latitude of 30 degrees and longitude 60 degrees. Where will it land on the surface of the Earth?

Chapter 2

UNIVERSAL LAW OF GRAVITATION

Contents

Newton discovered the law of gravitation through his studies of planetary motion. At the time of Newton, planetary motion was understood in terms of Kepler's three laws, which were based on the astronomical observations of Tycho Brahe. Newton was able to explain the origin of Kepler's laws from a more fundamental perspective by solving the equation of motion subject to a gravitational force given by his law of gravitation. Newton's successful explanation of the planetary laws of Kepler was a major achievement and led to the acceptance of his laws of mechanics and ushered in the new era of physics. We will study Kepler's laws of motion in the next section, and then, we will use Newton's laws of motion to derive them.

The Newtonian explanation of the planetary motion is based on the solution of the equations for the dynamics of one planet revolving around the Sun. You may recall that these problems are called two-body problems and we use the techniques of multi-particle systems to study them. It turns out that the motion of two bodies interacting with the gravitational force is an important problem that can be solved exactly and therefore serves as prototype for may other problems in physics. The dynamics of a three-body system, such as a system consisting of the Moon, the Earth and the Sun system, are too difficult to solve analytically, and the solutions of the equations of motion for these and more complicated systems have not been found yet. Much of the dynamical behaviors of the three-body system and more complicated systems are known through observations and numerical solutions.

2.1 KEPLER'S LAWS OF PLANETARY MOTION

Johannes Kepler (1571-1630) was an assistant to the great observational astronomer Tycho Brahe (1546-1601) who meticulously recorded the positions of planets for over twenty years at the observatory built on the island of Hveen with the financial support of the Danish king Frederik II. After Brahe's death in 1601, Kepler persuaded Brahe's widow to let him study the voluminous data collected by Tycho Brahe. Kepler originally believed that the motion of planets must be based on the perfect symmetry of the platonic solids and expected to find circular paths for the planetary motion. However, when he found that the observed orbit of Mars was not a circular motion, he had to abandon his belief, and instead, try to understand the data directly. With the new mindset, he discovered that planets follow an elliptic path rather than a circular path, which is called

Kepler's first law. He also deduced that the planets did not travel at constant speed around the ellipse, but the radius vector from the Sun to the planet sweeps out equal area in equal amount of time. This is called Kepler's second law. The first two laws were discovered in 1609. It took nine more years for Kepler to figure out the relation between the orbits of different planets, as formulated in the law of harmonies. The first two laws were not accepted for several decades, but the scientific community adopted the third law almost immediately after its publication in 1618. The three laws are as follows.

First Law or the Law of Elliptical Orbits - Planets travel in elliptical orbits about the Sun with the Sun at one focus.

Second Law or the Law of Equal Areas - The line joining the Sun and a planet covers equal area in equal time. Thus when a planet is nearer to the Sun, it has a higher speed than when it is further out. Figure 2.1 illustrates this law.

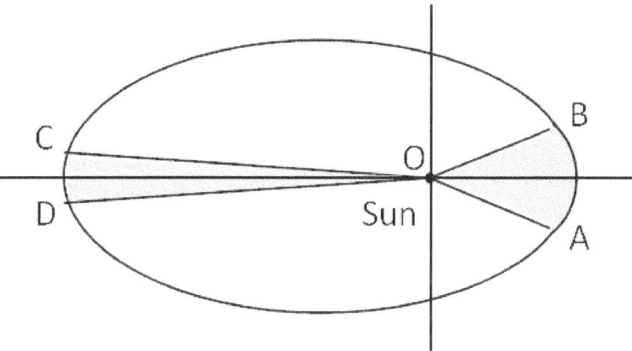

Fig. 2.1: The area OAB covered by a planet in a duration Δt is equal to the area OCD covered in the same duration Δt.

Third Law or the Law of Harmonies - The ratio of square of the period of revolution about the Sun to the cube of the semi major axis of the elliptical orbit of two planets are equal to each other. Thus if T_1 and T_2 are periods, and a_1 and a_2 the semi-major axes of two planets, we have the following equality.

$$\frac{T_1^2}{a_1^3} = \frac{T_2^2}{a_2^3} \tag{2.1}$$

Kepler's three laws have played critical role in the discovery of the universal law of gravitation. By 1666 Newton had early versions of his three laws of motion, but yet did not have the law of gravitation. In that year he had an insight that Earth's gravity extended also to the Moon and was counterbalanced by the centrifugal force. Newton used the balance of the two forces to find that the gravitational force must decrease as inverse square of the distance. The calculation was, however, only for a circular motion.

Newton did not work on the planetary motion problem any further till 1679 when another physicist, Robert Hooke, which you have met in the Hooke's law, went to see him about the elliptical orbit problem of the planets. Hooke had conjectured that a planet moving in an ellipse must be acted on by a central force by the Sun. Hooke had also come to the conclusion that this force must vary as the square of the inverse of the distance of the planet from the Sun, but he could not prove his conjectures mathematically.

After Hooke's visit, Newton went back to work on the planetary motion problem. First he showed mathematically that, if a body obeys Kepler's second law, then the force on the body must be central. Isaac Newton also showed that the angular momentum of a body is conserved if the body is acted upon by a central force. This finding demonstrated the physical basis of Kepler's second law as we have seen in the chapter on rotation.

Next, Newton showed that if a body is in an elliptical path, then the force must be pointed towards one of the foci and must vary as the square of the inverse of the distance from the focus. However, Newton did not publish any of these results until after the great Astronomer, Edmund Halley (1656-1742), asked him in 1684 if he could prove Hooke's conjecture. To Haley's surprise, Newton immediately replied that he had proved it five years earlier. Had Hooke been successful at proving his conjecture mathematically, we might be calling the law of gravitation the Hooke's law of gravitation instead of Newton's law of gravitation.

Halley used Newtonian calculations to ascertain that comets appearing in the sky in 1531, 1607 and 1682 were the same object and it was regularly appearing every 76 years. Sure enough the comet, now called Halley's comet, arrived on Christmas day in 1758 as predicted by Edmund Halley. The last time Halley's comet observable on Earth was in 1986 shown in Fig. 2.2. Upon Halley's insistence Newton wrote up his treatise on mechanics and its applications to the celestial mechanics called the Philosophiae naturalis principia mathematica, or *Principia*, as commonly known, which was published in 1687. You have already encountered a mention of this seminal work of Isaac Newton in the chapter on Newton's laws of motion.

Fig. 2.2: Halley's Comet. Picture by NASA (1986).

2.2 THE UNIVERSAL LAW OF GRAVITATION

In *Principia*, Newton presented his three laws of motion, and also stated the law of gravitation. Newton stated that any two particles in the universe exert an attractive force on one another whose magnitude is directly proportional to their masses m_1 and m_2 and decreases with the distance between the two masses as the square of the direct distance r between them.

$$\boxed{\text{Magnitude: } \quad F = G_N \frac{m_1 m_2}{r^2}.} \qquad (2.2)$$

The direction of the force on each mass is based on the attractive nature of the force - the force on m_1 is towards m_2 and on m_2 it is towards m_1. Here G_N is a constant, called the universal gravitational constant, which has the same value for the force between any two bodies. The constant G_N is considered to be truly universal, being same between any two objects anywhere in the universe. Do not confuse this G_N with acceleration due to gravity g. The value of G_N has been determined experimentally to be approximately 6.67428×10^{-11} m^3/kg.s^2.

Law for Spherical Bodies

Although Newton's law of gravitation stated in Eq. 2.2 is for point particles, the same form of the law applies to spherical objects provided we use the center-to-center distance for r. An important consequence of the inverse-square distance nature of the gravitational force is that, if you drill a hole inside a solid sphere of radius r_2 and place a mass m some distance r_1 from the center as shown in Fig. 2.3, then the force on mass will come from only the mass of the sphere that is inside the radius $r < a$, and none of the masses between radius

Further Remarks: The great English physicist Paul A. M. Dirac has hypothesized that the value of G_N may have been different in the past. A non-constant G_N will have significant cosmological consequences, which is not fully understood yet.

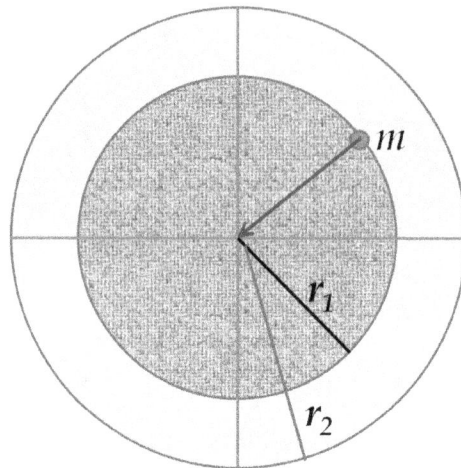

Fig. 2.3: The net gravitational force on a mass m placed at a point inside a spherical body is equal to the gravitational force of the particle of the sphere that is within the distance r_1 of the center of the sphere.

One of the consequences of this result is that if you sit inside a spherical shell, the gravitational force on you will be zero no matter where you sit inside the shell! Can you prove this assertion?

$r = r_1$ and $r = r_2$ will count if the sphere has uniform density. You can show it mathematically that the gravitational force on the test mass m from the particles of the sphere for whom the radial distance r is greater than r_1, i.e. $r_1 < r \leq r_2$, cancels out.

Law for Arbitrary Bodies

If the distance between two bodies is large compared to their sizes, the bodies can be replaced by point masses and then we can use the same form of the law as given in Eq. 2.2 for point particles. The force of attraction between non-spherical objects which are not far apart compared to their sizes is complicated and can only be computed by dividing the bodies into smaller parts and vectorially summing up the forces between every pair of masses.

Example 2.1. Newton's law of gravitation from a falling apple and falling Moon.

It is interesting to note that a straightforward analysis of falling of an apple and the continuous falling of the Moon towards the Earth can lead one to the discovery of the universal law of gravitation. Although, Newton did not come upon the universal law of gravitation in this way, the analysis is none-the-less instructive, and will be presented here.

From the observations on freely falling objects near the Earth, you know that the apple falls with an acceleration of 9.81 m/s². From observations on the Moon, i.e. from the distance to the Moon and its time period, it is also known that the Moon has a centripetal acceleration of 0.00272 m/s² in its circular orbit about Earth. Thus, the acceleration of the Moon is approximately 3600 times less than that of apple.

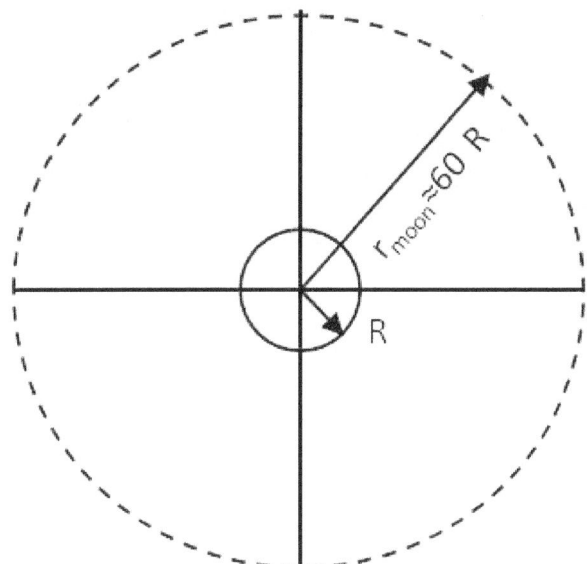

Fig. 2.4: The Moon traveling in circular motion about the Earth completes one orbit in approximately 29 days, which gives it's a centripetal acceleration of 0.00272 m/s².

$$\frac{a_{Moon}}{a_{apple}} = \frac{0.00272 \ m/s^2}{9.81 \ m/s^2} \approx \frac{1}{3600}$$

Since the Moon is at a distance of approximately 60 times the radius of the Earth, it can be concluded that the acceleration due to the gravitation of the Earth drops as the inverse of the square of the distance.

$$\frac{a_{Moon}}{a_{apple}} = \frac{r^2_{apple}}{r^2_{Moon}}$$

Since, force on an object is proportional to the acceleration, the force must also vary as $1/r^2$ if the acceleration varies as such. The genius of Newton was in the realization that the same law of force must apply between the Earth and the Moon as applies between the Earth and the apple.

$$\frac{F}{m} \propto \frac{1}{r^2}$$

Here m stands for the mass of the apple or the Moon.

Example 2.2. Gravitational force between objects

Find the gravitational forces between objects in the following situations. (a) Two spherical lead balls of masses 10 kg each separated by center-to-center distance of 10 cm. (b) Two people of height approximately 1.5 m and masses 100 kg and 60 kg separated by a distance of 2 m. (c) Same two people 20 m apart. (d) A satellite of mass 1000 kg in orbit 200 km above the Earth; use mass of the Earth = 5.96×10^{24} kg, and mean radius of the Earth = 6.37×10^6 m.

Solution. (a) We have stated above that a spherical object "acts" same as a point particle as far as gravitation outside the sphere is concerned. Therefore, the force between the spherical lead balls can be easily computed giving the following magnitude.

$$F = G_N \frac{m_1 m_2}{r^2} = \left(6.67 \times 10^{-11} \frac{\text{N.m}^2}{\text{kg}^2}\right) \times \frac{10 \text{ kg} \times 10 \text{ kg}}{(0.1 \text{ m})^2} = 6.67 \times 10^{-7} \text{ N}.$$

(b) Since people are not spherical objects, and since their separation is not too great compared to their sizes, we cannot use the same form of the law of gravitation as given for the point particles. If we use the formula given, we will only get some rough estimate, not an exact answer for the magnitude of the force.

$$F \approx G_N \frac{m_1 m_2}{r^2} = \left(6.67 \times 10^{-11} \frac{\text{N.m}^2}{\text{kg}^2}\right) \times \frac{100 \text{ kg} \times 100 \text{ kg}}{(2 \text{ m})^2} = 1.0 \times 10^{-8} \text{ N}.$$

(c) In this part, the people are separated by more than 10-times their sizes. Therefore approximating them as spheres will not give us too different a result than an exact result. Although the result of spherical approximation for a person will not give an exact result, but we expect the result to be more precise here than it was in part (b). The magnitude of the force is

$$F \approx G_N \frac{m_1 m_2}{r^2} = \left(6.67 \times 10^{-11} \frac{\text{N.m}^2}{\text{kg}^2}\right) \times \frac{100 \text{ kg} \times 100 \text{ kg}}{(20 \text{ m})^2} = 1.0 \times 10^{-10} \text{ N}.$$

(d) Although, the satellite is not a spherical object, it is far away from the Earth compared to the size of the satellite. Therefore, we can treat the satellite as a point mass. This gives the following for the magnitude of the force on the satellite.

$$F = \left(6.67 \times 10^{-11} \frac{\text{N.m}^2}{\text{kg}^2}\right) \times \frac{5.97 \times 10^{24} \text{ kg} \times 1000 \text{ kg}}{(6.37 \times 10^6 \text{ m} + 0.200 \times 10^6 \text{ m})^2} = 9.2 \times 10^3 \text{ N}.$$

Example 2.3. Gravitational force and circular orbits

A satellite is to be placed in a geocentric circular orbit about Earth. Find the altitude of the circular orbit above Earth.

Solution. A **geocentric orbit** means that the satellite has the same time period as the time period of the rotation of Earth. When a satellite is in a geocentric orbit, the satellite appears at the same spot above the Earth. These satellites are used for telecommunications. Let h be the altitude above Earth, and R_E the radius of the Earth. Then, the radius of the circular motion of the satellite is $R_E + h$. Since the satellite covers a distance of one circumference $2\pi(R_E + h)$ in time $T = 1$ day, the average speed of the satellite must be

$$v = \frac{2\pi(R_E + h)}{T}$$

We will assume that the speed of the satellite is constant. This means that the satellite would be in a uniform circular motion and the magnitude of the centripetal acceleration of the satellite would be

$$a_c = \frac{v^2}{R} = \left[\frac{2\pi(R_E + h)}{T}\right]^2 \frac{1}{R_E + h} = \frac{4\pi^2(R_E + h)}{T^2}$$

The centripetal acceleration is the net acceleration of the satellite in this case. The net force on the satellite is from the force from the Earth. Therefore, we can write the following using the magnitudes of the force on the satellite and the acceleration of the satellite.

$$ma_c = F \implies m\frac{4\pi^2(R_E + h)}{T^2} \approx G_N\frac{mM_E}{(R_E + h)^2}$$

We can solve this equation for the altitude of the satellite in a geocentric orbit with the following result.

$$h = \left(\frac{G_N M_E T^2}{4\pi^2}\right)^{1/3} - R_E = 3.6 \times 10^7 \text{ m}.$$

2.3 GRAVITATIONAL POTENTIAL ENERGY

Since the gravitational force is a conservative force, the work by this force can be expressed as a change in the potential energy of the object the force is acting. We have already derived the formula for the gravitational potential energy in the chapter on energy, Ch. **??**. Here, we present the derivation once again.

To find the expression for the gravitational potential energy we will calculate the work needed from an applied force $\vec{F}_{\text{appl}}$ to pull masses M and m apart from an initial distance of r_1 to the final distance of r_2. The work done gives the potential energy difference between the two states of the two-mass system: state (1) when the two masses are a distance r_1 apart and the state (2) when the two masses are a distance r_2.

$$\Delta U = -\int_i^f \vec{F}_{\text{appl}} \cdot d\vec{r} = \int_{r_1}^{r_2} \frac{G_N Mm}{r^2}dr = -G_N Mm\left(\frac{1}{r_2} - \frac{1}{r_1}\right)$$

In the chapter on energy, we had also defined a potential energy function by introducing a reference state. The potential energy of a system in an arbitrary state was given with respect to the potential energy in reference state taken as zero.

In the present case, the usual practice is to use the state when the two masses are infinitely apart since then they would have zero interaction. Therefore, we set $r_1 = \infty$, and drop the subscript from r_2 to obtain the formula for the gravitational potential energy of two masses m and M when they are an arbitrary distance r apart.

$$U(r) = -\frac{G_N M m}{r}.$$
(2.3)

Note that gravitational potential energy is not defined when the two bodies are on top of each other, i.e., when their separation is zero. We do not worry about this mathematical problem since such a configuration is physically impossible.

The gravitational potential energy gives the potential energy of the two masses together. That is, Eq. 2.3 is not the potential energy of m or M. Rather, it is the potential energy in the state of the two masses separated by a distance r. We will use potential energy function in applying conservation of energy to various problems.

The total energy of the two masses interacting with each other through a gravitational force only will be obtained by adding the kinetic energies of the two masses and the gravitational potential energy given above. That is, if the mass M has a speed V and the mass m a speed v at the time they are separated by a distance r, the total energy E of the two will be

$$E = \frac{1}{2}MV^2 + \frac{1}{2}mv^2 - \frac{G_N M m}{r}.$$
(2.4)

When the two masses move so that their separation distance changes, their speeds must also change so that remains unchanged. Note that if we were working only on mass m, then the energy of mass m will be

$$E_m = \frac{1}{2}mv^2 - \frac{G_N M m}{r} \text{(Energy of } m \text{ only)}.$$

And, when we are working with only mass M, the energy of M will be

$$E_M = \frac{1}{2}MV^2 - \frac{G_N M m}{r} \text{(Energy of } M \text{ only)}.$$

But their combined energy is as given in Eq. 2.4. The energy of the combined system is not equal to the sum of the energy of individual parts since the parts interact with each other and you must not count the energy of the same interaction twice.

Example 2.4. Placing a satellite in a geocentric orbit about the Earth. *A satellite of mass 1000 kg is placed in a geocentric orbit* *at an altitude of approximately 30,000 km from Earth. How much energy was expended to accomplish that? Mass of Earth $= 5.97 \times 10^{24}$ kg.*

Solution. The energy for sending satellites above the Earth is usually supplied by burning fuel as we have discussed in the chapter on impulse and momentum. The energy goes in lifting the material of the satellite from the surface of the Earth to the orbit. The burning of the fuel gives some initial kinetic energy to the satellite. We assume that the satellite starts out with some kinetic energy at the surface of the Earth which is converted to the greater potential energy when the satellite is in the orbit. We will calculate the required kinetic energy at the surface of the Earth as a measure of the energy needed to send the satellite to the orbit. Let us label quantities at the surface of the Earth by a subscript 1 and the corresponding quantities when the satellite is in the orbit by 2. Then, the required energy is

$$W_{\text{needed}} = K_1.$$

Now, from the conservation of energy of the satellite, we have

$$K_1 + U_1 = K_2 + U_2.$$

Therefore, we have

$$K_1 = K_2 + U_2 - U_1 = \frac{1}{2}mv_2^2 - \frac{G_N Mm}{r_2} + \frac{G_N Mm}{r_1}$$

From the given information in the problem, we can obtain the change in the potential energy, but we do not have the information for v_2 given in the problem. We can obtain v_2 from the equation of motion of the satellite when it is in the circular orbit. From the equation of motion of the satellite in a circular orbit at radius r_2 we have

$$m\frac{v_2^2}{r_2} = \frac{G_N Mm}{r_2^2}$$

Therefore, the energy equation becomes

$$K_1 = -\frac{G_N Mm}{2r_2} + \frac{G_N Mm}{r_1}$$

Now, we are ready to put in the numbers, and obtain the numerical answer.

$$W_{\text{needed}} = \left(6.67 \times 10^{-11} \frac{\text{N.m}^2}{\text{kg}^2}\right) \times 5.97 \times 10^{24} \text{ kg} \times 1000 \text{ kg} \times$$
$$\left(\frac{1}{6.37 \times 10^6 \text{m}} - \frac{1}{2(6.37 \times 10^6 \text{m} + 30 \times 10^6 \text{m})}\right)$$
$$= 5.7 \times 10^{10} \text{ J}.$$

How does this energy compare to the chemical energy in gasoline? Googling for the energy in gasoline we find that one kilogram of conventional gasoline contains approximately 4.4×10^7 J. Therefore, we would need energy in approximately 750 kg gasoline to put a 1000-kg satellite into the geosynchronous orbit 30,000 kg above the surface of Earth.

2.4 THE TWO-BODY PROBLEM FOR GRAVITATIONAL FORCE

2.4.1 CM and Relative Position Coordinates

In this section we will revisit the two-body problem discussed earlier in the book. This time we will discuss the problem in the context of a gravitational force between two bodies. Consider an isolated system consisting of two particles of masses m_1 and m_2 interacting with a gravitational force. Let $\vec{r}_1$ and $\vec{r}_2$ be the position vectors of particles 1 and 2 respectively with respect to an inertial frame $Oxyz$ as shown in Fig. 2.5.

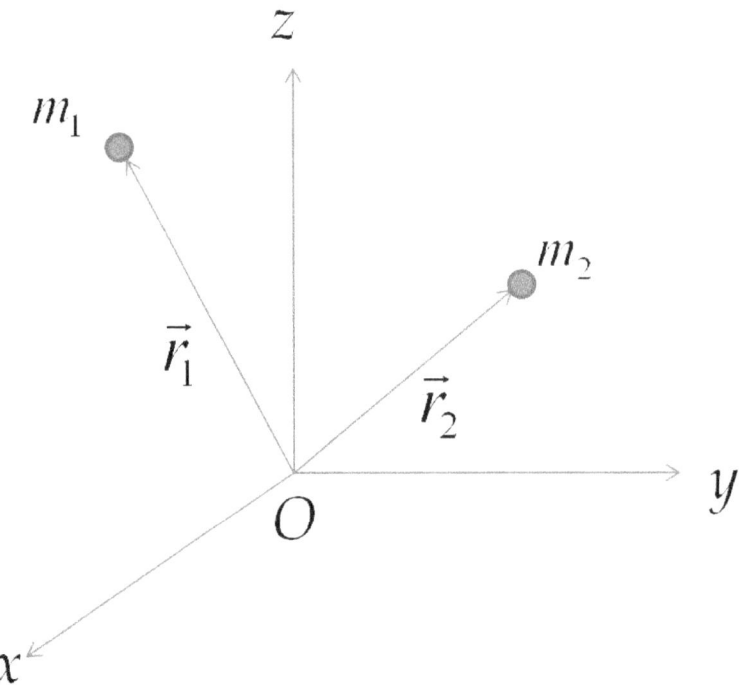

Fig. 2.5: Position vectors of two objects of a two-body system in an inertial frame.

Here, the only force on either of the body is the gravitational force by the other body. The force on m_1 is directed from 1 to 2, and on m_2, it is directed in the opposite direction. The force on m_1 written in the vector notation takes the following form.

$$\vec{F_1} = -\frac{G_N m_1 m_2}{r^2} \hat{u}_{2\to1} \tag{2.5}$$

where $\hat{u}_{2\to1}$ is a unit vector from position of m_2 towards the position of m_1. Therefore, the equations of motion for the six coordinates of two bodies, viz., $\vec{r_1} = (x_1, y_1, z_1)$ and $\vec{r_2} = (x_2, y_2, z_2)$ are as follows.

$$m_1 \frac{d^2 \vec{r_1}}{dt^2} = -\frac{G_N m_1 m_2}{r^2} \hat{u}_{2\to1} \tag{2.6}$$

$$m_2 \frac{d^2 \vec{r_2}}{dt^2} = +\frac{G_N m_1 m_2}{r^2} \hat{u}_{2\to1} \tag{2.7}$$

There is an alternate set of six coordinates that is easier to work with than the set of Cartesian coordinates of the two bodies. They consist of the three coordinates of the center of mass $\vec{R}$ and three for the relative position vector $\vec{r}$ shown in Fig. 2.6.

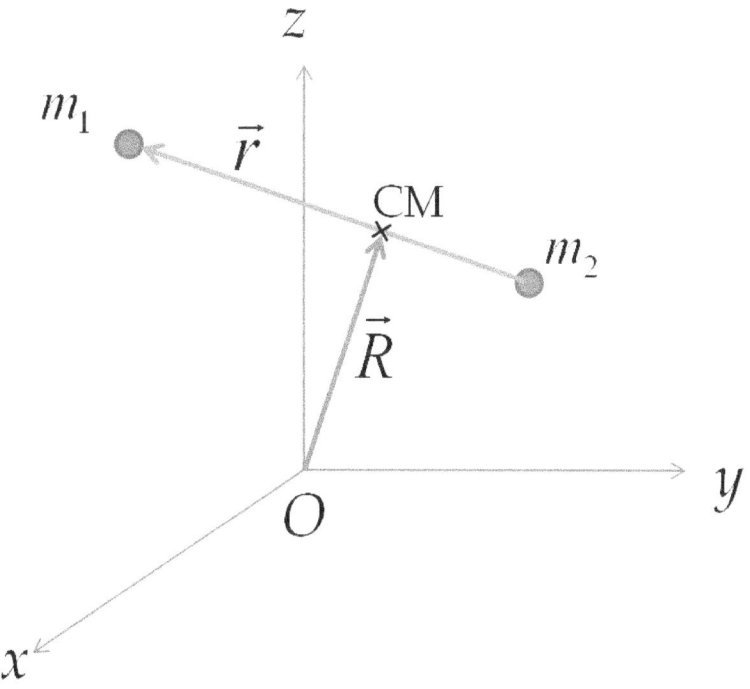

Fig. 2.6: Position vectors of two objects of a two-body system.

The CM and the relative coordinates are defined as follows in terms of the Cartesian coordinates $\vec{r_1}$ and $\vec{r_2}$ of the two particles.

$$\vec{R} = \frac{m_1 \vec{r}_1 + m_2 \vec{r}_2}{m_1 + m_2} \tag{2.8}$$

$$\vec{r} = \vec{r}_1 - \vec{r}_2 \tag{2.9}$$

The relative position vector $\vec{r}$, defined this way, gives the position of the particle 1 with respect to the position of the particle 2. The equations of motion for these abstract coordinates can be obtained by taking time derivative of the $\vec{R}$ and $\vec{r}$ vectors, and then using the equations of motions for the position coordinates $\vec{r}_1$ and $\vec{r}_2$ of the two particles based on Newton's second law of motion given in Eqs. 2.6 and 2.7. After some algebra which I leave to the student to complete, we find that the following equations of motion for the position of the center of mass and the relative position of particle 1 with respect to particle 2.

$$\boxed{\frac{d^2 \vec{R}}{dt^2} = 0} \tag{2.10}$$

$$\boxed{\frac{d^2 \vec{r}}{dt^2} = -\frac{G_N(m_1 + m_2)}{r^2} \hat{u}_{2 \to 1}} \tag{2.11}$$

These equations show that the motion of the CM is decoupled from the motion about the CM, similar to the decoupling of translation and rotational motion we have seen before. The acceleration of the CM of a two-body system in an inertial frame is zero, as expected, because the two bodies form an isolated system whose total momentum is conserved.

The position of m_1 with respect to m_2, given by $\vec{r}$, changes in a way that it is possible to recast the problem in terms of one hypothetical body. Let us divide both sides of Eq. 2.11 by $m_1 + m_2$, and multiply the result by $m_1 m_2$.

$$\left(\frac{m_1 m_2}{m_1 + m_2}\right) \frac{d^2 \vec{r}}{dt^2} = -\frac{G_N m_1 m_2}{r^2} \hat{u}_{2 \to 1} \tag{2.12}$$

The quantity in parenthesis on the left side of the equation has the units of mass, and is called the **reduced mass**. The reduced mass is often denoted by the Greek letter μ (pronounced mu).

$$\boxed{\mu = \frac{m_1 m_2}{m_1 + m_2} \quad \Longleftrightarrow \quad \frac{1}{\mu} = \frac{1}{m_1} + \frac{1}{m_2}.} \tag{2.13}$$

Note that when one of the masses is much greater than the other, the reduced mass is very close to the smaller mass.

$$\mu \approx m_2 \quad \text{if} \quad m_1 >> m_2.$$

Let us write M for the total mass $m_1 + m_2$.

$$M = m_1 + m_2.$$

We also note that the unit vector $\hat{u}_{2 \to 1}$ from 2 to 1 is a unit vector in the direction of the relative position vector $\vec{r}$. Therefore, for simpler notation, we use $\hat{u}_r$ for this unit vector.

$$\hat{u}_r \equiv \hat{u}_{2 \to 1}.$$

Now, the equation of motion of the relative position can be written in terms of the total mass and the reduced mass in the following form, which "looks like" the equation of a single particle of mass μ in the gravitational force of another particle of mass M.

$$\boxed{\mu \frac{d^2 \vec{r}}{dt^2} = -\frac{G_N M \mu}{r^2} \hat{u}_r.} \qquad (2.14)$$

We say that the original equations of motion Eqs. 2.6 and 2.7, in which the motions of the particles m_1 and m_2 are coupled, has been decoupled into motions of two fictitious particles:

One of these fictitious particles has a mass M and moves at constant velocity with respect to the origin of the inertial frame. The second fictitious particle has a mass μ and moves in the gravitational field of another particle of mass M which is fixed at the origin.

These two de-coupled variables are easier to work than the original coordinates. Therefore, we will focus solving then and deducing the original coordinates from the solutions of $\vec{R}$ and $\vec{r}$.

2.4.2 Equations of Motion in the CM Frame

Since there are no external forces on an isolated two-body system, the CM does not accelerate. Therefore, it is possible to choose our inertial frame fixed to the CM. In that case $\vec{R}$ will be zero, $\vec{R} = 0$ (CM frame), and will remain zero for all time. The original particles and the relative position coordinates are shown in Fig. 2.7. Using the definition of CM in terms of the coordinates of the individual objects in the CM frame, we find the following relation between the coordinates of the two bodies in the CM frame.

$$\vec{R} = \frac{m_1 \vec{r}_1\,' + m_2 \vec{r}_2\,'}{m_1 + m_2} = 0 \implies m_1 \vec{r}_1\,' + m_2 \vec{r}_2\,' = 0. \qquad (2.15)$$

The relative position vector of the two masses in the CM-frame is also equal to $\vec{r}$ since a displacement vector is independent of the choice of origin.

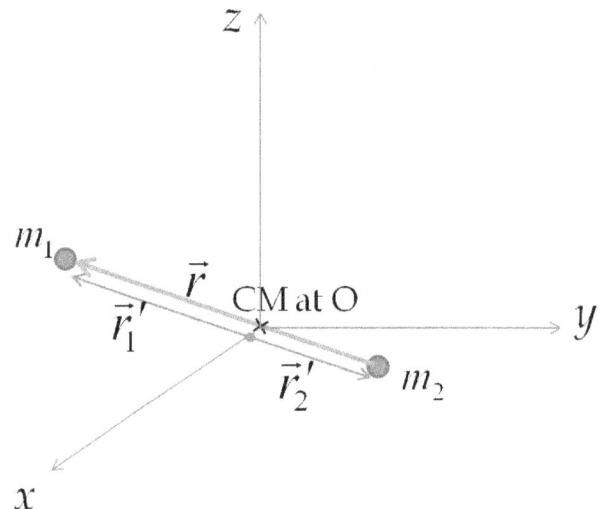

Fig. 2.7: The CM frame. The positions of m_1 and m_2 in the CM frame are denoted by $\vec{r}_1'$ and $\vec{r}_2'$, respectively. The relative position of 2 with respect to 1, $\vec{r} = \vec{r}_1' - \vec{r}_2'$, which has the same formal appearance in this coordinate as it had in the inertial frame whose origin was at some other place than at the CM.

$$\vec{r} = \vec{r}_1 - \vec{r}_2 = \left(\vec{R} + \vec{r}_1'\right) - \left(\vec{R} + \vec{r}_2'\right) = \vec{r}_1' - \vec{r}_2'. \qquad (2.16)$$

Combining Eqs. 2.15 and 2.16 we find that if we work in the CM frame, we will be able to easily obtain the motion of each body from the motion of the relative coordinate only by the following relations.

$$\vec{r}_1' = \frac{m_2}{m_1 + m_2}\vec{r} = \left(\frac{\mu}{m_1}\right)\vec{r} \qquad (2.17)$$

$$\vec{r}_2' = \frac{m_1}{m_1 + m_2}\vec{r} = \left(\frac{\mu}{m_2}\right)\vec{r} \qquad (2.18)$$

If one of the masses, say m_1 is very much larger than m_2, i.e. $m_1 \gg m_2$, as is the case for instance for the Sun/Earth system, then the relative coordinate is simply the coordinate of the smaller mass with respect to the CM. We will make this assumption in the rest of the chapter.

When $m_1 \gg m_2$, then $\mu \approx m_1$, $\vec{r}_1' \approx 0$ and $\vec{r}\,'2 \approx \vec{r}$.

2.4.3 Angular Momentum Conservation and Kepler's Second Law

Further simplification of analysis of a two-body system takes place by exploiting the fact that for an isolated system, the total angular momentum is conserved. The angular momentum $\vec{L}$ of the two objects is the sum of their individual angular momenta. Writing the angular momentum about the CM we have

$$\vec{L} = \vec{r}_1' \times m_1 \vec{v}_1' + \vec{r}_2' \times m_2 \vec{v}_2'. \quad \text{[About the CM]} \qquad (2.19)$$

By a series of algebraic manipulations you can show that this equation simplifies to the following.

$$\boxed{\vec{L} = \vec{r} \times \mu \frac{d\vec{r}}{dt}. \quad \text{[About the CM]}} \qquad (2.20)$$

Therefore, the angular momentum of the two bodies about their CM is simply equal to the angular momentum of a hypothetical object of mass equal to the reduced mass and the position equal to the relative position of the two objects.

Since neither the magnitude nor the direction of the conserved angular momentum $\vec{L}$ can change with time, the motion of the effective body of mass μ must be in a plane perpendicular to the direction of $\vec{L}$ as illustrated in Fig. 2.8.

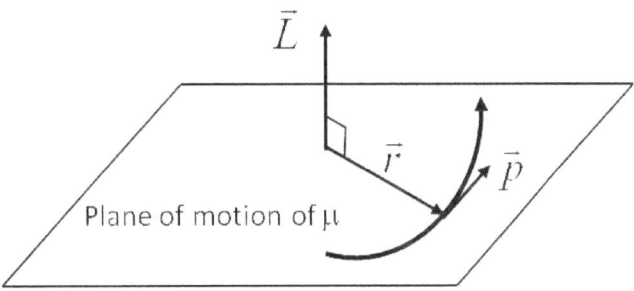

Fig. 2.8: The reduced mass moves in a plane.

Let us choose the coordinate system so that the z axis is pointed along the conserved angular momentum direction, and the position of reduced mass is measured from the origin of this coordinate system. Let l be the magnitude of the angular momentum. Then, the angular momentum vector of the system is written as

$$\vec{L} = l\hat{u}_z. \qquad (2.21)$$

The magnitude of the conserved angular momentum places the following constraint on the relative position.

$$\left| \vec{r} \times \mu \frac{d\vec{r}}{dt} \right| = l. \qquad (2.22)$$

Kepler's second law of equal areas is a particular consequence of the conservation of angular momentum as seen by expressing this equation in polar coordinates (r, θ). This calculation here is a repeat of a similar calculation we have done in the chapter on rotation.

$$\mu r^2 \left| \frac{d\theta}{dt} \right| = l. \tag{2.23}$$

Here both r and θ are functions of time. From Fig. 2.9, it is clear that the area covered by the orbit of the reduced mass μ in time Δt is $r^2 \Delta \theta / 2$. We can write the area covered in a duration in terms of the conserved angular momentum of the object.

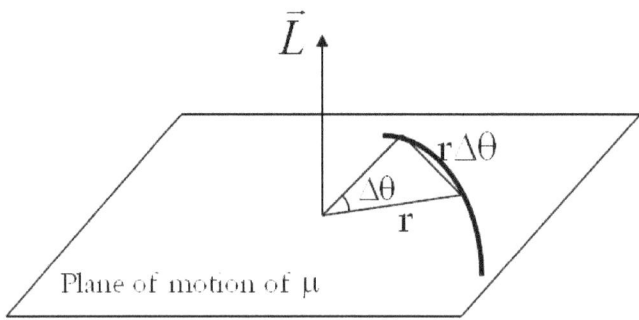

Fig. 2.9: Finding area spanned by the reduced mass in its orbit.

$$\text{Area covered during } \Delta t \equiv \Delta A = \frac{r^2 \Delta \theta}{2} = \frac{l}{2\mu} \Delta t.$$

Therefore, the rate at which area swept by a line from the origin to the position of the particle is

$$\frac{\Delta A}{\Delta t} = \frac{l}{2\mu}, \text{ constant.}$$

Therefore, we can say that the radial vector from the Sun to a planet sweeps out an equal area in an equal time. This is Kepler's second law of planetary motion which was originally obtained empirically from examining Tycho Brahe's data on the motion of the planet Mars. Thus, Kepler's second law of motion is simply a consequence of the conservation of angular momentum of the combined Sun/Planet system.

2.4.4 Energy Conservation

Since gravitational force is a conservative force, the total energy of the two-body interacting via the gravitational force is conserved. The total energy E of the two-body system is given by the following.

$$E = K + U = \left(\frac{1}{2}m_1 v_1^2 + \frac{1}{2}m_2 v_2^2 \right) - G_N \frac{m_1 m_2}{r} \qquad (2.24)$$

By using polar coordinates and substituting various quantities, you should show that this expression can also be written as follows.

$$\begin{aligned} E &= \frac{1}{2}\mu \left(\frac{dr}{dt} \right)^2 + \frac{1}{2}\mu r^2 \left(\frac{d\theta}{dt} \right)^2 - G_N \frac{m_1 m_2}{r} \\ &= \frac{1}{2}\mu \left(\frac{dr}{dt} \right)^2 + \left(\frac{1}{2}\frac{l^2}{\mu r^2} - G_N \frac{m_1 m_2}{r} \right) \end{aligned} \qquad (2.25)$$

In polar coordinates, we notice that the kinetic energy of the angular motion, i.e. the centrifugal energy, acts like an additional term for the potential energy. We combine the actual potential energy (the last term) with the second term on the right side in the above equation and define an effective potential energy $U_{eff}(r)$.

$$\boxed{U_{\text{eff}} = \frac{1}{2}\frac{l^2}{\mu r^2} - G_N \frac{m_1 m_2}{r}.} \qquad (2.26)$$

Let us now rewrite the energy of a planet of mass m moving in a space where it has only the gravitational force of a Sun of mass M as

$$\boxed{E = \frac{1}{2}\mu v_r^2 + U_{\text{eff}} \equiv K_r + U_{\text{eff}},} \qquad (2.27)$$

where K_r stands for the kinetic energy of the planet, given by $\frac{1}{2}\mu v_r^2$ with v_r for the radial speed of the planet. Note that K_r is not the kinetic energy of the planet since the angular part of the kinetic energy has been lumped with the potential energy to obtain the effective potential energy. We will see below that the accounting of energy in this modified way helps with the analysis of the motion of the planet.

This equation "looks like" an equation of energy of a particle of mass μ, not the mass m of the planet. This fictitious particle appears to move in a "one dimensional world" of radial coordinate, whose value ranges from zero to positive infinity. Of course, in the real world, the planet is moving in a plane with both the radial coordinate and the angular coordinate. But, as far as Eq. 2.27, both K_r and U_{eff} depend only on one coordinate r. Therefore, we may pretend that there is only one coordinate r to worry about. But, this coordinate is weird since it has only positive values, unlike the x, y, z-coordinates.

Interpreting the Effective Potential Energy

Now, we examine Eq. 2.27 for various possible values of energy E. Note that K_r cannot be negative in Eq. 2.27. Therefore, $E - U_{\text{eff}}$ must always be greater than or equal to zero.

$$E - U_{\text{eff}}(r) \geq 0, \quad \text{since } K_r \geq 0, \tag{2.28}$$

where I have indicated the dependence of U_{eff} on r explicitly. For given masses M, m and magnitude of the angular momentum l, the effective potential takes a definite functional form. To explore the implications of these inequalities, it is helpful to write the equality in Eq. 2.28, viz. $E = U_{\text{eff}}(r)$ more explicitly using the expression for $U_{\text{eff}}(r)$. We will also write $U_{\text{eff}}(r)$ as $a/r^2 - b/r$ for simplicity.

$$U_{\text{eff}} = \frac{1}{2}\frac{l^2}{\mu r^2} - G_N \frac{m_1 m_2}{r} \Leftrightarrow a/r^2 - b/r.$$

Equation $E = U_{\text{eff}}(r)$ is then a quadratic equation in r.

$$Er^2 + br - a = 0.$$

For $E = 0$ this equation has the solution

$$r = \frac{a}{b} = \frac{l^2}{2G_N m_1^2 m_2^2} \quad [E = 0].$$

and for non-zero E, the solution is

$$r = \frac{1}{E}\left(-b \pm \sqrt{b^2 + 4aE}\right) \quad [E \neq 0]. \tag{2.29}$$

If $E > 0$, then we see that the radical in this solution gives a larger value than b. Therefore, only plus of $\pm$ will be physical since $r > 0$ here. This gives one physical solution for $E > 0$.

$$r = \frac{1}{E}\left(-b + \sqrt{b^2 - 4aE}\right), \quad [E > 0]. \tag{2.30}$$

If $E < 0$, then, the radical will be real if $b^2 - 4a|E| > 0$. Let us write the solution with absolute value of energy $|E|$ so that we have easy time making sure $r > 0$ in the final analysis.

$$r = \frac{1}{|E|}\left(b \mp \sqrt{b^2 - 4a|E|}\right). \tag{2.31}$$

Now, we note that since $b^2 - 4a|E| > 0$, we will have $b^2 > 4a|E| > 0$. Therefore, we will always have $b > \sqrt{b^2 - 4a|E|}$, which would give two values for r.

$$r_1 = \frac{1}{|E|}\left(b - \sqrt{b^2 - 4a|E|}\right) \quad [E < 0] \tag{2.32}$$

$$\text{and, } r_2 = \frac{1}{|E|}\left(b + \sqrt{b^2 - 4a|E|}\right) \quad [E < 0] \tag{2.33}$$

Finally, the negative value for E cannot be below the minimum of U_{eff}. This show up in the requirement of the reality of the radical in the solution when $E < 0$. We find that when $E < 0$, then

$$|E| \leq \frac{b^2}{4a}.$$

At the minimum, $|E| = \frac{b^2}{4a}$. Using this condition in the solution we find that r has only one value.

$$r = \frac{2a}{b} \quad E = E_{\min}. \tag{2.34}$$

A plot of U_{eff} shown in Fig. 2.10 helps visualize these different types of solutions. In this plot the four values of interest of the total energy of the system are indicated on the ordinate axis. The four solutions have the following meaning.

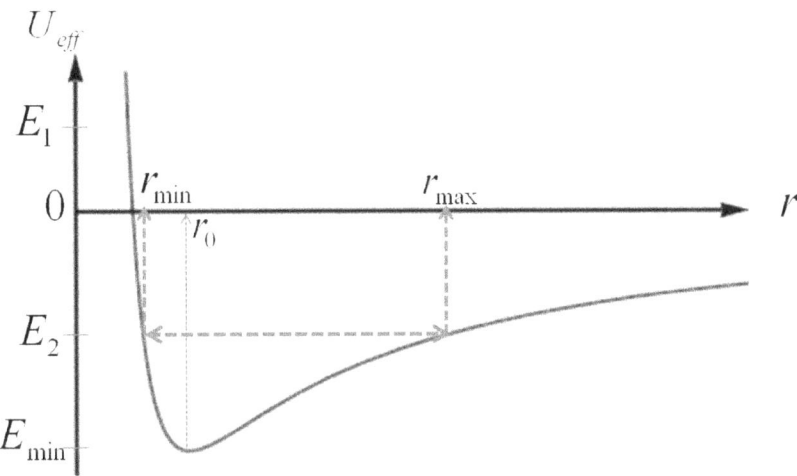

Fig. 2.10: Plot of effective energy versus radial distance r. Four energy levels of interest are indicated on the axis for U_{eff}. They are $E = 0, E_1, E_2, E_{\min}$.

Case 1: $E = 0$. There is only one radial distance when $E = U_{\text{eff}}$. At this point K_r must be zero. Below we will look at the equation of the trajectory and find that the trajectory of μ is parabolic. That is, the particle can move to a closest distance to the origin where the "force-exerting" fictitious particle of mass M is located. The position where $K_r = 0$ corresponds to $v_r = 0$, which is the point where the radial velocity will change from being positive to being negative and vice-versa. This point in space is called the turning point of the radial coordinate. Note that the radial velocity zero does not mean that the full velocity of the particle is zero there, since we also have angular velocity, which would be non-zero.

The picture of the original two-particle would be that the two particles approach one another to a closest approach and then return to be far apart. They are kept apart due to the centrifugal energy barrier. The motion is an unbounded one.

Case 3: $E > 0$. This case is similar to the case of $E = 0$ in the sense that there is only one turning point at the solution. The two bodies approach one another to a closest approach and then return to be far apart. They are kept apart due to the centrifugal energy barrier. The motion is an unbounded one. Although, this case appears to be similar to $E = 0$ case, but we will find below that the trajectory is not parabolic but has the shape of a hyperbola.

Case 3: $U_{eff}^{min} < E < 0$. In this case, there are two turning points corresponding to the two solutions, the smaller value of the solution is called r_{min}, or closest approach, and the larger one r_{max}, or furthest apart. In the Sun/Earth system, the r_{min} and r_{max} are called **perihelion** and **aphelion**. In the Earth/Satellite system, these points are called **perigee** and **apogee**. The motion is bounded with a maximum and a minimum separation between the two bodies. A rigorous calculation shows that the trajectory is an ellipse.

Case 4: $E = U_{eff}^{min}$. This is a particular sub-case of case 3, and corresponds to a circular motion as is evident in only one solution to the condition $E = U_{eff}$ or $K_r = 0$. The radial component of the velocity is zero at all points of the trajectory, i.e. the velocity has only angular component in this case.

2.4.5 The Orbit Equation and Kepler's First Law

We saw above that the orbits of two bodies interacting with a gravitational force is more readily presented in the CM-frame. In the CM frame, it suffices to work out the equivalent problem of the orbit of a fictitious particle with mass equal to the reduced mass. Once, the orbit of the fictitious particle is known, we can immediately work out the orbits of two masses by the relation between the coordinates of the reduced mass and the coordinates of the individual masses.

Often, it is not necessary to work out the orbits of the individual masses since we often deal with systems where one of the bodies has much more mass then the other. For instance, in the Sun/Earth system, the mass of the Sun is much more than the mass of Earth. In these systems, we approximate the motion of the smaller body by the motion of the reduced mass and represent the larger mass as a point mass at the CM, where the origin is also placed.

If the masses of the two bodies are equal to each other or close to each other, as maybe the case of two stars in a binary star system, we cannot make the above simplification. The motion of the fictitious particle in these cases does not approximate to the motion of either body. We must deduce the motion of each mass either directly or indirectly. However, even in these cases, we find that the equation of motion of the reduced mass is very helpful in deducing the motion of each mass. We work out the motion of the reduced mass and deduce the motion of each mass indirectly from the solution.

Therefore, regardless of whether we are dealing with disparate mass system or closer mass system, it is always a better strategy to think in terms of the reduced mass. We will examine the motion of the reduced mass in this section more fully. We will continue to call the particle with the reduced mass a fictitious particle just to emphasize the fact that reduced mass is a deduced quantity, the actual particles being m_1 and m_2.

To obtain the orbit of the fictitious particle μ in a two-body system it is sufficient to examine the conservation of angular momentum and the conservation of energy in the system. We will place the origin at the CM and write the energy and the angular momentum conservation equations in the polar coordinates as follows.

$$\mu r^2 \frac{d\theta}{dt} = l, \text{ constant angular momentum} \tag{2.35}$$

$$\frac{1}{2}\mu \left(\frac{dr}{dt}\right)^2 + \frac{1}{2}\frac{l^2}{\mu r^2} - \frac{G_N m_1 m_2}{r} = E, \text{ constant energy} \tag{2.36}$$

Note that, although the angular momentum and the kinetic energy of the two bodies in the system can be written completely in terms of the one-body corresponding to the reduced mass particle, the potential energy still contains the original masses m_1 and m_2 as parameters in the equations.

The equations corresponding to the conservation of energy and angular momentum can be solved to gain information about the types of orbits possible for given l, E, m_1 and m_2. The equation of the orbit is obtained by dividing dr/dt by $d\theta/dt$ so that we can get an equation for $dr/d\theta$.

$$\frac{dr}{d\theta} = \frac{dr/dt}{dr/d\theta} = \frac{\text{From Eq. 2.36}}{\text{From Eq. 2.35}}. \tag{2.37}$$

This equation can be integrated to find radial coordinate as a function of the angular coordinate, $r(\theta)$, for the orbit. The calculus needed to perform these operations is somewhat beyond this course. If you are intrigued, you will have to look up a more advanced book on mechanics to find the necessary algebra, or better yet, carry out the calculations yourself. I will state the final answer here so that we can discuss the physical consequences of the result, which are important for understanding the motion of planets. The solution $r(\theta)$ is as follows.

$$\boxed{r = \frac{r_0}{1 - e\cos\theta}} \qquad (2.38)$$

where

$$r_0 = \frac{l^2}{G_N m_1 m_2 \mu} \qquad (2.39)$$

$$e = \sqrt{1 + \frac{2El^2}{G_N^2 m_1^2 m_2^2 \mu}} \qquad (2.40)$$

The parameter r_0 corresponds to the radius of the circular orbit and e is called the eccentricity of the orbit, which characterizes the shape of the orbit. To identify the orbits as ellipses, circles, parabolas or hyperbolas, it is better to write the orbit equation in more familiar Cartesian coordinates. Using $x = r\cos\theta$ and $y = r\sin\theta$ we find the following equivalent form of the orbit equation.

$$\left(1 - e^2\right) x^2 - 2r_0 e x + y^2 = r_0^2. \qquad (2.41)$$

Now, we can see how the four cases of different types of orbits depend on energies of the system, through the parameter e. To discuss the type of orbits that result for various values of energy as compared to the minimum effective potential energy, it is best to rewrite the parameters r_0 and e in terms of the minimum value of the effective potential energy $U_{\text{eff}}^{\min}$, which we will denote by symbol U_0. By an elementary calculation, you can show that the minimum value of the effective potential energy is

$$U_0 = -\frac{(G_N m_1 m_2)^2 \mu}{2l^2} \qquad (2.42)$$

Therefore, the parameters r_0 and e in the orbit equation are

$$r_0 = \frac{1}{\sqrt{-2\mu U_0}} \qquad (2.43)$$

$$e = \sqrt{1 - \frac{E}{U_0}}. \qquad (2.44)$$

Case 1 $E = 0$. If $E = 0$, then $e = 1$. The coefficient of x^2 in Eq. 2.41 vanishes which leads to a parabola.

Case 2 $E > 0$. If $E > 0$, then $e > 1$ since $U_0 < 0$. This will give an opposite sign to the coefficient of x^2 than that of the y^2. Therefore, this is an equation of a hyperbola. The fictitious particle μ has a closest approach to the CM given in the figure by the point when the hyperbolic orbit crosses the x-axis. (Fig. 2.11). Setting $y = 0$ in the orbit equation, we solve for x, and choose the smaller of the two negative x where the closest approach occurs.

Case 3 $U_0 < E < 0$. This case corresponds to $0 < e < 1$. Both x^2 and y^2 have the same sign and unequal coefficients. Hence, the

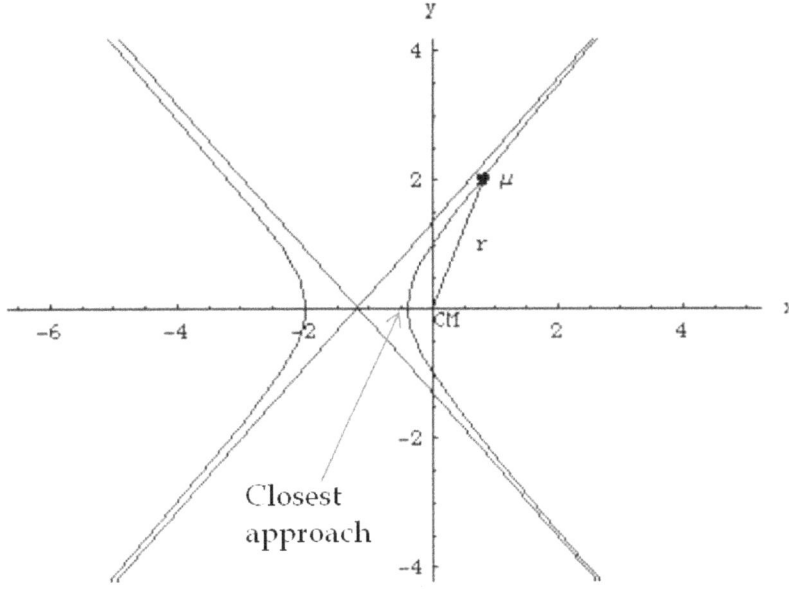

Fig. 2.11: An elliptic orbit for $r_0 = 1$ unit and $e = 1.5$ for an attractive potential. The other branch is for a repulsive force since acceleration will be pointed away from the CM.

orbit is an ellipse. Using the Cartesian form of the orbit equation, the semi-major axis a and semi-minor axis b are given by

$$a = \left| \frac{r_0}{1 - e^2} \right| = \frac{G_N m_1 m_2}{2|E|} \tag{2.45}$$

$$b = \frac{r_0}{\sqrt{1 - e^2}} = \frac{l}{\sqrt{2\mu|E|}} \tag{2.46}$$

The closest and the farthest approaches occur on the semi-major axis, which is along the x-axis here, as shown in Fig. 2.12. The farthest distance r_{max} is obtained by setting $\theta = 0$ in the orbit equation.

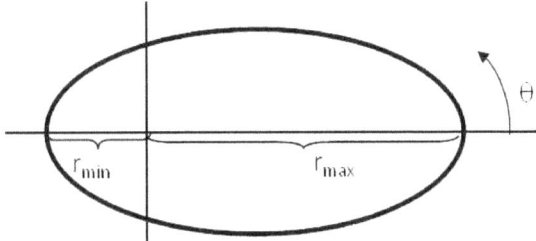

Fig. 2.12: Closest and farthest approach of a object in elliptical orbit.

$$r_{max} = r|_{\theta=0} = \frac{r_0}{1 - e} \tag{2.47}$$

The closest distance occurs for $\theta = 180°$.

$$r_{min} = r|_{\theta=180°} = \frac{r_0}{1 + e} \tag{2.48}$$

Of course, you could also obtain the closest and farthest distances from the CM by setting $y = 0$ in the Cartesian form of the orbit equations, and solving the quadratic equation for x. These are the perihelion and aphelion or perigee and apogee points for planets and satellites

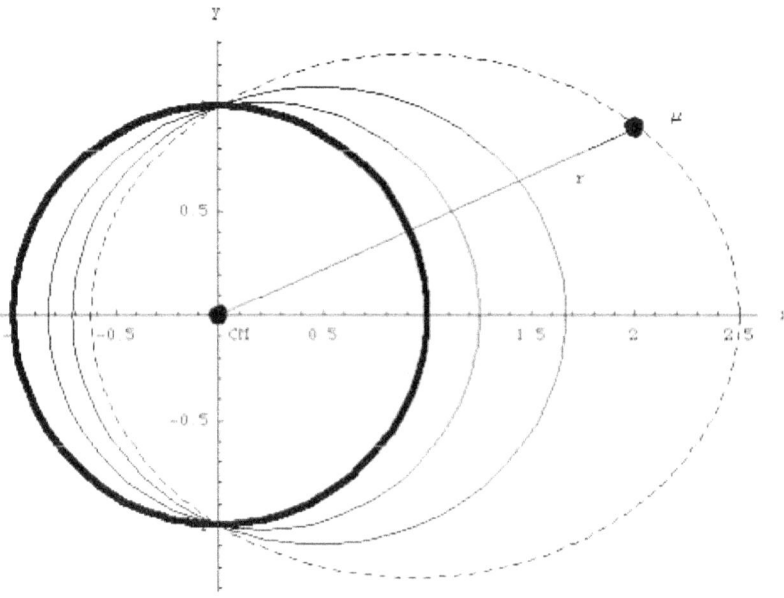

Fig. 2.13: Elliptical orbits for three different eccentricities 0.2, 0.4, 0.6. The thick solid line is a circular orbit which corresponds to $e = 0$. The higher the eccentricity more elongated the ellipse. The value of r_0 was set to 1 for the plot.

Table 2.1: Eccentricities of solar planets and planet-like objects

Planet	Eccentricity (e)
Mercury	0.206
Venus	0.007
Earth	0.017
Mars	0.093
Jupiter	0.048
Saturn	0.055
Uranus	0.051
Neptune	0.007
Pluto	0.252
Sedna	0.8

Case 4 $E = U_0$. This makes $e = 0$. With $e = 0$, the orbit equation becomes $x^2 + y^2 = r_0^2$, an equation of a circle of radius r_0.

The eccentricity of orbits tells us about the elongation of the ellipse; the larger the eccentricity the more elongated the ellipse. Thus when $e = 0$, ellipse is same as a circle, and when $e = 0.6$, the ellipse is elongated as shown in Fig. 2.13.

Both cases 3 and 4 correspond to bound orbits. Hence planets (as well as comets) have either elliptical or circular orbits around the Sun confirming Kepler's first law. The orbits of different planets differ in their eccentricities. The orbit of Venus is almost a circle while the orbit of Pluto is quite elongated. Comets have much higher eccentricities. For instance, the orbit of Halley's comet has an eccentricity of 0.967. Recall that if $e > 1$ the orbit will not be bounded, but hyperbolic. Table 2.1 summarizes eccentricities of planets. Except for Mercury and Pluto, all other planets have minor eccentricities.

2.4.6 Kepler's Third Law From the Orbit Equation

We can deduce Kepler's third law if we apply orbit equation to two planets revolving around the Sun. Rather than directly deal with the orbit equation, we first work out a relation between the period of a planet of mass m in orbit around the Sun of mass M and the planet's distance r from the Sun. We work with polar coordinates with the planet in xy plane and the angular momentum pointed along z axis. Let l be the z component of the angular momentum. We have seen above that

$$l = \mu r^2 \frac{d\theta}{dt}$$

Here μ is the reduced mass of m and M. Rearrange the equation, and divide both sides by 2.

$$\frac{l}{2\mu} dt = \frac{1}{2} r(rd\theta)$$

The right side is a differential element of area enclosed in the ellipse centered about the focus. Therefore, upon integration over a whole period we will obtain the area of the ellipse, πab, on the right side. On the left side integration will give a quantity proportional to the period T.

$$\frac{l}{2\mu} T = \pi ab \tag{2.49}$$

From the formulas for a and b given above, we can express b in terms of a.

$$b^2 = \left(\frac{l^2}{G_N M m \mu} \right) a,$$

where the quantities within parenthesis are constants. Squaring both sides of Eq. 2.49, rearranging terms, and putting b in terms of a we find the following for a planet of mass m around the Sun of mass M.

$$\frac{T^2}{a^3} = \frac{4\pi^2}{G_N(M + m)}. \tag{2.50}$$

Kepler's third law compares T^2/a^3 for two planets around the Sun. Their ratios will be

$$\boxed{\frac{(T^2/a^3)_1}{(T^2/a^3)_2} = \frac{M + m_2}{M + m_1}.} \tag{2.51}$$

Since planets' masses m_1 and m_2 are considerably smaller than the mass M of the Sun, we can expand $1/M + m_1$ in the Maclaurin series series and keep the leading terms to obtain

$$\frac{(T^2/a^3)_1}{(T^2/a^3)_2} = 1 + \frac{m_2 - m_1}{M} + \cdots \tag{2.52}$$

According to Kepler's third law the ratio of the square of the period to the cube of the semi-major axis of a planet's motion is independent of the planet, but we find here that T^2/a^3 does depend upon the masses m of the planets. Thus, **Kepler's third law is not an exact law**. The deviation can be expressed in terms of m/M of planets which we show in Table 2.2. Even for the most massive planet, Jupiter, m/M is of the order of 0.001% only.

Table 2.2: Ratio of Mass of Planets to Mass of Sun

Planet	m $(10^{23}$ kg)	m/M $\times 10^{-07}$
Mercury	3	1.51
Venus	50	25.1
Earth	60	30.2
Mars	6	3.02
Jupiter	20000	10100
Saturn	6000	3020
Uranus	900	452
Neptune	1000	503

Example 2.5. Europa around Jupiter. *Europa is a moon of Jupiter that is most like Earth. It has a mass of 4.8×10^{22} kg and moves in an ellipse of eccentricity 0.009 with a mean distance of $670,900$ km from the center of Jupiter whose mass is 1.9×10^{27} kg. What are the energy and the angular momentum of Europa? Ignore the motion around the Sun.*

Solution. Since the mass of Europa is so much smaller than that of Jupiter, we can assume one-particle picture and set μ equal to the mass m of Europa. The data is given for the average distance between Europa and Jupiter which we will equate to r_0 since if Europa moved at the average distance its distance will not change and the orbit would be circular.

$$e = 0.009$$
$$r_0 = 6.7 \times 10^8 \ m$$

We need $G_N M m$ to find energy E. We will calculate the numerical value of this quantity so that we do not clutter the final formulas.

$$G_N M m = 6.7 \times 10^{-11} \frac{\text{N.m}^2}{\text{kg}^2} \times 1.9 \times 10^{27} \text{kg} \times 4.8 \times 10^{22} \text{kg} = 6.1 \times 10^{39} \text{N.m}^2.$$

Hence, the energy of Europa will be

$$E = -\frac{G_N M m}{2a} \approx -\frac{G_N M m}{2r_0} = -4.6 \times 10^{30} \text{ J}.$$

We use the energy to find the magnitude of the angular momentum.

$$l = b\sqrt{-2\mu E} \approx r_0 \sqrt{-2\mu E} = 4.5 \times 10^{35} \text{ kg.m}^2/\text{s}.$$

The direction of the angular momentum is pointed perpendicular to the plane of the orbit.

Example 2.6. Satellite about the Earth

A satellite of mass 3000 kg is put in an elliptic orbit about the Earth with the semi-major and the semi-minor axes of the orbit being 10,000 km and 8,000 km respectively. (a) What are the energy and angular momentum of the satellite? (b) How far away is the satellite from the center of the Earth at its nearest approach, perigee, and its farthest point, apogee? (c) What are the speed of the satellite at the perigee and the apogee? Ignore the effect of the Sun.

Solution. Once again, we have the situation $m \ll M$. Therefore, we will substitute m for μ.

(a) We have $G_N M m = 1.2 \times 10^{18}$ N.m^2, $a = 1.0 \times 10^7$ m, and $b = 8.0 \times 10^6$ m. Therefore, the energy of the satellite is

$$E = -\frac{G_N M m}{2a} = -6.0 \times 10^{10} \text{ J},$$

and the angular momentum has the following magnitude.

$$l = b\sqrt{-2\mu E} = 1.5 \times 10^{14} \text{ kg.m}^2/\text{s}.$$

The direction of the angular momentum is pointed perpendicular to the plane of the orbit.

(b) We will use the orbit formula to answer these questions. The eccentricity is obtained from a and b.

$$e = \sqrt{1 - \frac{b^2}{a^2}} = 0.6.$$

The circular orbit radius is

$$r_0 = a(1 - e^2) = 6.4 \times 10^6 \text{ m}.$$

The perigee $r_{\min}$ occurs at $\theta = 180°$ and apogee $r_{\max}$ at $\theta = 0°$ in the orbit equation $r = r_0/(1 - e\cos\theta)$.

$$r_{\max} = 1.6 \times 10^7 \text{ m}$$
$$r_{\min} = 2a - r_{max} = 4.0 \times 10^6 \text{ m}$$

(c) Equating the energy E of the satellite to the sum of the kinetic and potential energies, we find the speed of the satellite.

$$E = \frac{1}{2}\mu v^2 + U \approx \frac{1}{2}mv^2 - \frac{G_N M m}{r} \implies v = \sqrt{\frac{2}{m}\left(E + \frac{G_N M m}{r}\right)}.$$

We use the value of energy from part (a) that is same for all r and obtain the value of speed at the perigee by setting $r = r_{\min}$ and at apogee by setting $r = r_{\max}$ in this equation. The values obtained are $v_{\text{perigee}} = 12,700$ m/s and $v_{\text{apogee}} = 3,200$ m/s.

2.5 TIDAL FORCES

When a uniform force, such as the approximately constant force of gravity near the Earth of magnitude (mg) acts on an object, each particle of the object accelerates equally. What happens when force varies from place to place, such as the universal force of gravitation. For instance, the force of gravitation on Earth from the Sun is different at different parts of Earth since different parts of the Earth are at different distances from the center of the Sun. The force on the side of the Earth that faces the Sun will be larger than the force on the opposite side of the Earth.

A force that varies with the location is called a non-uniform force. Gravitational and electrical force are common examples of non-uniform forces. When a body is subjected to a non-uniform force, different parts of the body experience different forces, and consequently accelerate to different extent. As a result,a deformable body under a non-uniform force will be deformed as illustrated in Fig. 2.14. On the other hand, if the object is not deformable, then stresses in the body will develop as a result of different forces in different parts of the body. If the stress inside the body is large enough, the body may break apart as a result of the non-uniform force.

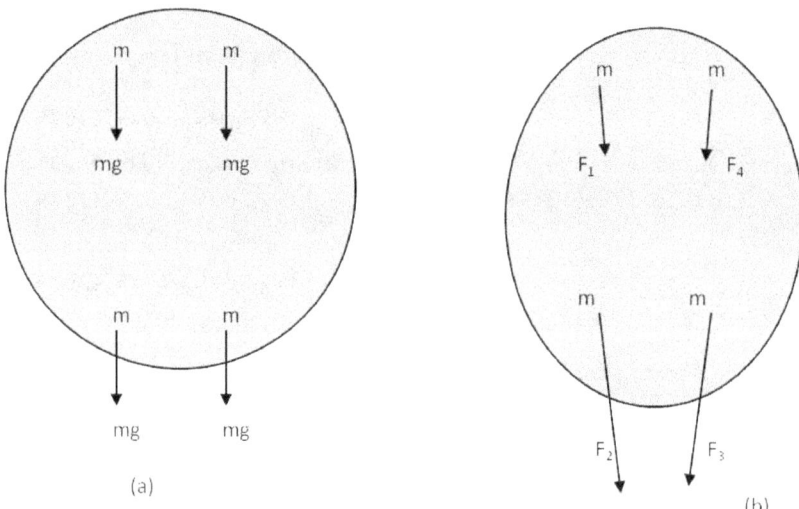

Fig. 2.14: (a) In a uniform gravity, equal forces act on equal masses m of at all parts of the body. (b) Under a non-uniform force, which is the case with the gravitational force, not the approximate force of gravity near the Earth, different parts of a body experience different force, which causes the body to stretch out in the direction of the force.

The deformation of a body or stress in a body subject to a non-uniform force is said to be caused by the differential force that depends on how the force varies over the body. The ocean tides on the

Earth are largely caused by the non-uniform gravitational force of the Moon as illustrated in Fig. 2.15.

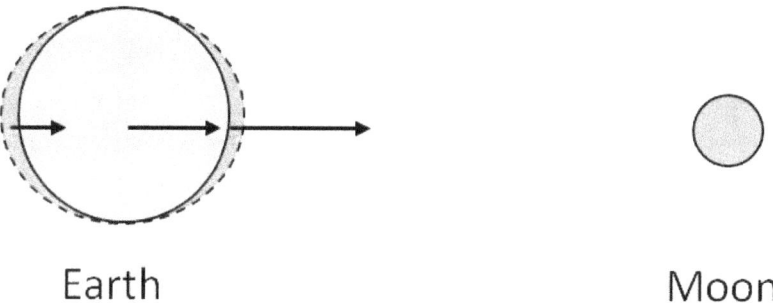

Fig. 2.15: The ocean tides on the Earth are primarily as a result of the tidal forces from the non-uniform gravitational force of the Moon. The ocean water on the side facing the Moon is pulled towards the Moon due to a stronger force on that side, and at the same time, the solid part of the Earth is pulled towards the Moon and away from the the ocean water on the far side. Every 24 hours, a side of the Earth is either facing towards the Moon or facing away from the Moon, and that is why we have two tides per day.

Due to this association with the ocean tides, the differential forces on a body due to a non-uniform force are also called the **tidal forces**. Although, the net force of the Sun on the Earth is far greater than the force of the Moon on the Earth, the gradient at which the force of the Moon varies over the Earth is greater than the gradient at which the force of Sun varies on the Earth due to the relative proximity of the Moon.

To understand the origin of a tidal forces, consider two equal masses $m_1 = m$ and $m_2 = m$ connected by a spring placed near a large mass M. Let the separation of the masses be $2a$ and distance from the center of M to the midway between the two masses be r as shown in Fig. 2.16.

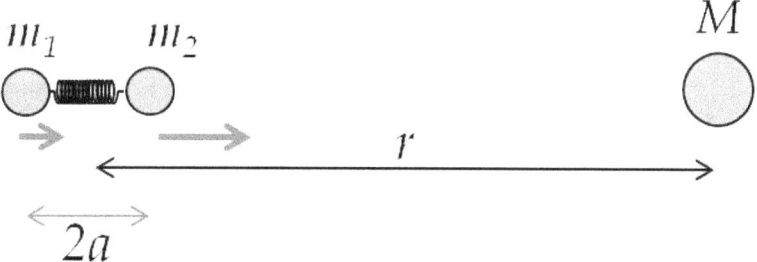

Fig. 2.16: A system of two masses near another mass will have a tidal force on the two mass system.

The differential tidal force on the two-mass system will be the difference of the forces on the two masses by the external body M.

$$|\Delta\vec{F}| = GMm \left[\frac{1}{(r-a)^2} - \frac{1}{(r+a)^2} \right]$$

Often, we are interested in situations when $a << r$. We can try to expand the quantity in the parenthesis in a Maclaurin series for a/r, and keep the first non-vanishing term.

$$\frac{1}{(r \pm a)^2} = \frac{1}{r^2}\left(1 \pm \frac{a}{r}\right)^{-2} = \frac{1}{r^2}\left(1 \mp \frac{2a}{r} + \cdots\right)$$

Therefore,

$$\frac{1}{(r-a)^2} - \frac{1}{(r+a)^2} = \frac{1}{r^2}\left(1 + \frac{2a}{r} + \cdots\right) - \frac{1}{r^2}\left(1 - \frac{2a}{r} + \cdots\right)$$

$$\approx \frac{4a}{r^3}, \text{ only the leading term.}$$

Keeping only the leading non-vanishing term yields the following expression for the tidal force in case of $a << r$.

$$\boxed{|\Delta\vec{F}| = \frac{4GMma}{r^3}.}$$

Therefore, the tidal force goes as $1/r^3$, which explains why the tidal force by the Moon on the Earth is much larger than that by the Sun, even though the later is much more massive.

The tidal force basically acts as a net repulsive force within the body along the direction of the non-uniform force due to a larger acceleration of one side of the body as compared to the other side. Therefore, if the tidal force is greater than the binding force of the body, the object will break apart.

Naturally, the tidal forces have enormous consequences in Astronomy since the forces in large object dealt in the Astronomy are considerable compared to the binding forces among materials. Astronomical objects, such as stars, planets, moons, and comets are held together by gravitational attraction. For instance, when a two-body system held together by their mutual gravitational attraction comes close to another massive body, the two bodies can be torn apart due to the tidal force from the massive body on the two-body system. There is a critical distance, called the **Roche limit** after the French physicist Edouard Roche, for this to happen.

The Roche limit of a two mass-system can be worked out by demanding that the tidal force from the external large mass must be greater than the force of attraction between the two masses. Let the two-mass system have equal masses m and be separated by a distance $2a$. Let the distance between the center of the two-mass system to the center of an external body of mass M be r as shown in Fig. 2.16. Then, Roche limit will require that

$$|\Delta\vec{F}|_{\text{tidal}} > F_{\text{between m's}} \quad (\text{break up condition})$$
$$\implies \frac{4GMma}{r^3} > \frac{Gmm}{4a^2}$$
$$\implies r < a\left[\frac{16M}{m}\right]^{1/3}.$$

The idea of disintegration of the two-body system shown here can be extended to the disintegration of a single body such as the Earth due to the tidal force from an external body such as the Sun. If the Sun came closer than the Roche limit, also called the Roche radius, the Earth will disintegrate. Sure enough, you don't find any moons within the Roche limits of any planets. The bodies that revolving around the planets, Jupiter, Saturn, Uranus and Neptune within their Roche limits are small rocks forming rings unable to clump into larger masses.

2.6 EXERCISES

Gravitational Force

Ex 2.1. Evaluate the magnitude of gravitational force between (a) two 5-kg spherical steel balls separated by a center-to-center distance of 15 cm, (b) the Sun and the planet Neptune which have an average separation of 30.15 AU. $M_{Sun} \approx 2.0 \times 10^30 \ kg$ and $M_{Neptune} \approx 1.0 \times 10^26 \ kg$.

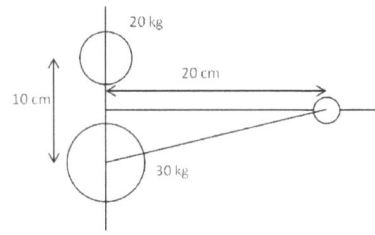

Fig. 2.17: Problem 2.2.

Ex 2.2. Find the net gravitational force on a 2-kg steel spherical ball by two other spherical balls with masses 20-kg and 30-kg in the configuration shown in Fig. 2.17. Give both the magnitude and the direction.

Ex 2.3. Find the gravitational force of the Moon on a bucket of water of mass 1 kg placed at two sides of the Earth, one facing the Moon and the other exactly on the opposite side. (a) Find the force of the Sun on the two buckets. (b) Think of a way to use your calculations to explain why there are two tides per day.

Ex 2.4. There is a small sphere of mass m from the edge of a uniform rod of length L and mass M as shown in the Fig. 2.18. Find the gravitational force on the mass m. [Hint: You will need to integrate.]

Fig. 2.18: Problem 2.4.

Ex 2.5. There is a small sphere of mass m from the edge of a uniform rod of length L and mass M as shown in the Fig. 2.19. Find the gravitational force on the mass m.

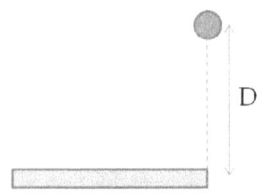

Fig. 2.19: Problem 2.5.

Ex 2.6. There is a small sphere of mass m at a height D from the center of a uniform ring of radius R and mass M as shown in the Fig. 2.20. Find the gravitational force on the mass m.

Ex 2.7. Evaluate the gravitational potential energy between (a) two 5-kg spherical steel balls separated by a center-to-center distance of 15 cm, (b) the Sun and the planet Neptune which have an average separation of 30.15 AU. $M_{Sun} \approx 2.0 \times 10^30$ kg and $M_{Neptune} \approx 1.0 \times 10^26$ kg.

Ex 2.8. Find the escape speed of a projectile from the following planets, (a) Earth, (b) Mars, (c) Saturn, and (d) Jupiter.

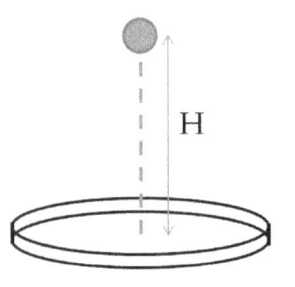

Fig. 2.20: Problem 2.6.

Ex 2.9. (a) From the average distance of the Earth from the Sun, and the orbital period of the Earth, find the centripetal acceleration of the Earth for its motion about the Sun. (b) Compare the value found in (a) to the gravitational force of the Sun on the Earth per unit mass of the Earth.

Ex 2.10. (a) A satellite is in a circular orbit around the Earth at a distance R from the center of the Earth. Find a formula for its period in terms of the distance R, the mass M of the Earth and the radius R_E of Earth. (b) Find an expression for the mechanical energy of the satellite.

Ex 2.11. (a) Find the reduced mass of the two-body system, called the Earth/Sun system consisting of the Earth and the Sun. (b) Where is the center of mass of the Earth/Sun system located. (c) Find the angular momentum of the Earth about the center of mass of the Earth/Sun system. (d) Where in the orbit does the Earth move fastest and where does it move slowest? Why? (e) Is the kinetic energy of the Earth with respect to the CM of the Earth/Sun system constant during its flight around the Sun? (f) Is the angular momentum of the Earth with respect to the CM of the Earth/Sun system constant during its flight around the Sun? (g) Write an energy based on the energy of the Earth/Sun system that is conserved. (h) Write an angular momentum based on the angular momentum of the Earth/Sun system that is conserved.

Ex 2.12. Derive $\vec{r}_1{}'$ and $\vec{r}_2{}'$ in terms of the relative coordinate $\vec{r}$ in the CM frame.

Ex 2.13. Derive $\vec{L} = \vec{r} \times \mu \frac{d\vec{r}}{dt}$ from $\vec{L} = \vec{r}_1{}' \times m_1\vec{v}_1{}' + \vec{r}_2{}' \times m_2\vec{v}_2{}'$.

Kepler's Laws and Orbit Equation

Ex 2.14. The closest and farthest distance of the Earth from the Sun are 0.98 AU and 1.02 AU, where, the Astronomical Unit, 1 AU = 149,598,000 km. Find (a) the semi-major axis and (b) the eccentricity of Earth's orbit.

Ex 2.15. The closest and farthest distance of Mars from the Sun are 1.38 AU and 1.67 AU. Find (a) the semi-major axis, (b) the eccentricity , and (c) the period of Mars's orbit.

Ex 2.16. The closest and farthest distance of Mercury from the Sun are 0.31 AU and 0.47 AU. Find (a) the semi-major axis, (b) the eccentricity , and (c) the period of Mars's orbit.

Ex 2.17. The orbit of Halley's comet is approximately elliptical with $e = 0.967$. Halley's comet comes around every 76 years. Find (a) the distance of the closest approach to the Sun and (b) the farthest distance the comet goes from the Sun.

Ex 2.18. A satellite of mass 2000 kg is put in an elliptical orbit of eccentricity 0.5 about Mars. The distance from the center of the planet to the closest approach of the satellite is equal to $\frac{7}{6}$ times the radius of the planet. (a) Find the distance to the farthest point. (b) Find the energy and angular momentum of the satellite. (c) Find the speed at the closest approach. (d) Using the angular momentum conservation between the farthest and the closest approach, find the speed at the farthest point.

Ex 2.19. What is the radius of the circular orbit for a satellite of mass 2500 kg about the Earth if it has an energy of -2×10^9 J?

2.7 PROBLEMS

P 2.1. Prove that the gravitational force on a point particle of mass m placed outside a spherical shell of mass M is equal to the gravitational force on m by a point mass of mass M placed at the center of the shell.

P 2.2. Prove that the gravitational force on a point particle of mass m inside a spherical shell of mass M is zero.

P 2.3. A tunnel is dug through the center of Earth. Show that a particle of mass m dropped in the tunnel will execute a simple harmonic motion, and deduce the frequency of oscillation of m.

P 2.4. A planet of mass m moves around a star of mass M. For an inertial observer the planet and the star appear to move in circles of radii r and R respectively. For $m = M/4$ and $r = 1.0 \times 10^{10}$ m, find radius R of the orbit of the star.

P 2.5. A satellite of mass 1000 kg is in circular orbit about Earth. The radius of the orbit of the satellite is equal to two times the radius of Earth. (a) How far away is the satellite? (b) Find the mechanical energy of the satellite. (c) Find the angular momentum of the satellite?

P 2.6. A spaceship of mass 3000 kg is to be sent from Earth to Venus. Assume the orbits of the Earth and the Venus around the Sun to be

approximately circular. (a) What is the minimum energy required
for the transfer? (b) The most efficient transfer from one circular
orbit to another circular orbit about the Sun is achieved by what is
called **Hohmann transfer**, which uses an elliptical orbit around the
Sun as an intermediary with the smaller and larger circles located at
perihelion and aphelion of an ellipse as shown in Fig. 2.21. The speed

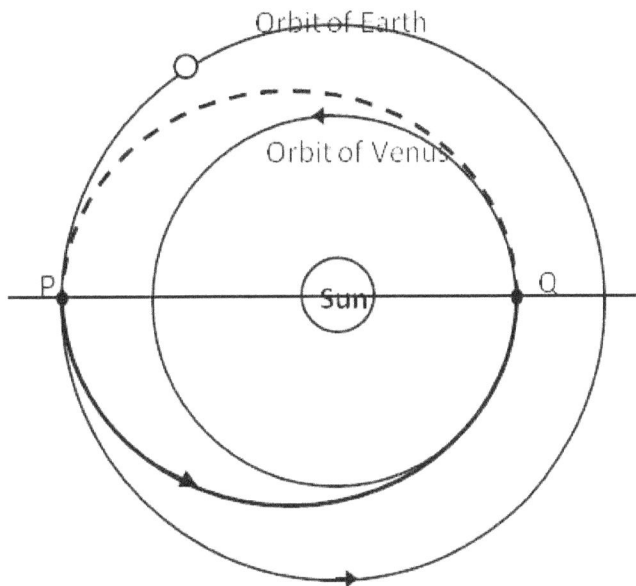

Fig. 2.21: Problem 2.6.

of the satellite must be increased at point Q and reduced at point
P for the transfer. Find the changes in speed required at points P
and Q. Note that the spaceship will have speeds of Earth and Venus
while in their orbits. Ignore the gravitational pulls of the Earth and
the Venus on the spaceship.

P 2.7. A satellite is in an elliptical orbit about the Earth with a
minimum altitude of 1500 km and a maximum altitude of 8000 km. In
one of the flights, its engine is fired when it is at the closest approach
point so that its speed increases by 10%. You can assume the speed
increased over a very short period of time. What is the maximum
altitude reached in its new elliptical orbit? Ans: 3.14×10^7 m.

P 2.8. Two objects of masses m and M interact with a central force
that varies as $1/r^4$ with proportionality constant k. Derive a formula
for the potential energy function. State the location of the reference
for your formula of potential energy?

P 2.9. A space station of mass m moves in a circular orbit of radius R
around Jupiter (mass M). (a) A rocket of mass Δm is fired radially

inward towards the center of Jupiter from the space station which causes the space station to "instantaneously" acquire a radial velocity v_r in addition to the angular velocity v_θ it had before the rocket firing. As a result the space station is thrown into an elliptical orbit. Find the semimajor axis and eccentricity of the elliptical orbit. (b) What would have happened if the rocket was fired tangentially to the circular orbit with relative speed u towards the front?

Chapter 3

SIMPLE HARMONIC MOTION

Contents

Oscillatory motion is ubiquitous is nature. You find the oscillatory motions in the swaying of branches of trees, rocking of boats, the back and forth motion of pendulums, vibrations of atoms in molecules, vibrations of a guitar string, etc, just to name a few. The oscillatory motion of varied objects share common characteristics that can be understood from studying a simple system of a block attached to a spring. We will study this model system in detail in this chapter.

3.1 SIMPLE HARMONIC OSCILLATIONS

3.1.1 Oscillatory Motion

Consider a block attached to a spring and placed on a frictionless table as shown in Fig. 3.1. The other end of the spring is attached

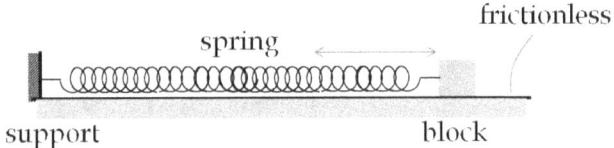

Fig. 3.1: A spring between a block and a fixed support provides interaction between the block and the support. The spring force on the block causes the block to oscillate back and forth.

to a fixed support. You might say that the spring is the vehicle of interaction between the block and the support. The forces of gravity and normal on the block in the vertical direction are balanced.

Note that when the spring is in the relaxed state, there is no net force between the block and the support. This condition is called static equilibrium. The block is said to be in an **equilibrium** and the position of the block in this situation is called the **equilibrium position** of the block.

Restoring Force

Suppose, you pull the block a little distance from the equilibrium position away from the support, and release from rest, the block would begin to oscillate. The oscillations of the block will be accompanied by the successive contraction and expansion of the spring. The same thing happens if you push the block towards the support and let go of it. In either case, there would be a force on the block when it is away from the equilibrium position depending on the stretch or the contraction of the spring. The direction of this force is always towards the equilibrium position regardless of where the block is in its cycle. This type of force is called a **restoring force** since this force attempts to bring the object back to the equilibrium position.

Let us look at the direction of the force on the block at various points in one cycle of the motion of the block. Suppose the cycle of motion is $AOBOA$ in Fig. 3.2. When the block is at A, the force on the block is pointed to the left towards the equilibrium position O. When the block is at the equilibrium point O, there is no net force on the block. When the block is at B, which is on the other side of the equilibrium from point A, the force is pointed to the right towards O.

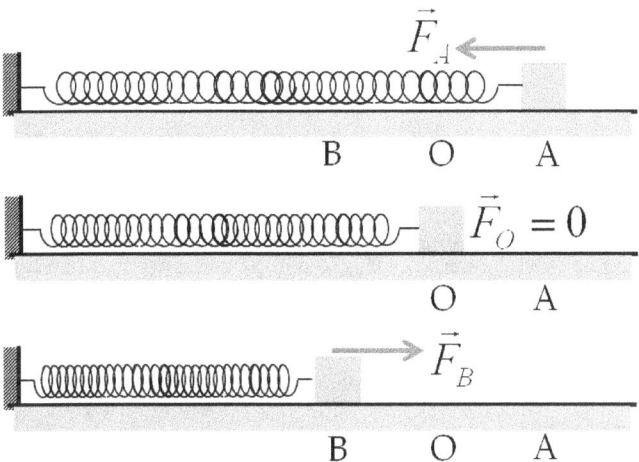

Fig. 3.2: The force on the block changes direction as the block moves in one cycle about the equilibrium point O. The force on the block is always pointed towards the equilibrium point, and when the block is at the equilibrium point, the force is zero. The force on the block is labeled with the label of the point in the motion of the block.

Velocity Versus Acceleration

Since the direction of the acceleration is always towards the equilibrium point and the velocity is in the direction of motion, the acceleration and velocity will not be in the same direction at all points of the cycle of the oscillation. For instance, when the block is moving away from the equilibrium position, the velocity of the block would be pointed away from the equilibrium point but its acceleration will be pointed towards the equilibrium. Therefore, when the block is moving away from the equilibrium, the block will slow down and come to a stop and reverse its direction of motion as shown in Fig. 3.3.

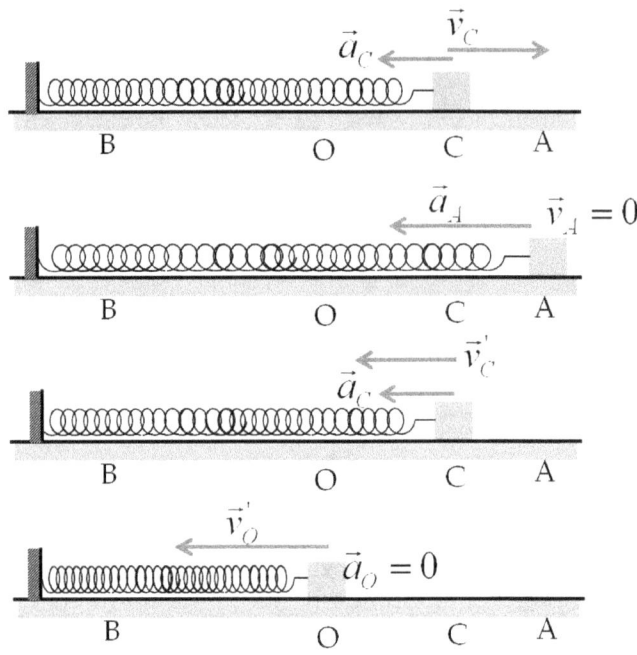

Fig. 3.3: A part of the cycle of motion of a block attached to a spring is shown here. In the first figure at the top, the block is moving towards the right at point C and the acceleration is pointed to the left. This slows the block and brings the block to rest at point A. The acceleration of the block at A points in the direction of AO, i.e. towards the equilibrium O. When the block returns to point C, it is now moving with the velocity in the same direction as the acceleration. Therefore, the block continues to speed up and reaches the equilibrium point with the largest speed in the cycle. at the equilibrium the acceleration is zero. But, when the block goes on the other side of the equilibrium point, the velocity would be pointed to the left and the acceleration will be pointed towards the right. That would slow the block and would bring it to rest at point B. Now, due to the non-zero acceleration pointed to the right, the speed of the block will increase till it reaches back to the equilibrium point. Points A and B are the turning points of the motion of the block.

At other points, when the block is moving towards the equilibrium, the velocity and acceleration of the block would be in the same direction. Therefore, when the block is moving towards the equilibrium, it will speed up. Suppose the block is moving towards the equilibrium from the right, say in the AOB direction in figure. As shown in the figure, when the block reaches the equilibrium point, its velocity will be pointed to the left and the acceleration will be zero at that point. The block will then continue to move to the left of the equilibrium. But, at those points, the acceleration would be pointed to the right which is in the opposite direction to the velocity at O at this instant. That means that, the block has the largest speed at the moment it passes through the equilibrium point O since after that the speed will start to decrease. The block would come to rest again at the other turning point B. After that, the block will pick up speed towards O, since during B to O, both the velocity and accelerations would be in the same direction.

The points where the velocity turns around are called **turning points**. In the present case, there will be two turning points, one to the right of O and one to the left of O. The block will move between these two turning points. The distance between the equilibrium point and the turning points is called the **amplitude** of the motion.

Thus, the block may start with zero velocity at a turning point, pick up speed, reach maximum speed at the equilibrium, then slow down to rest at the other turning point, then pick up speed again, reach maximum speed at the equilibrium, then slow down to rest at the first turning point. Therefore, a block attached to a spring is capable of back and forth motion indefinitely. This back and forth motion between the turning points is also called an **oscillatory motion**.

Further Remarks:

The oscillatory motion illustrated for a block attached to a spring is possible for many other systems, such as pendulum, bridge, buildings and towers. In all these physical situations, we notice that a back and forth oscillatory motion is possible due to the presence of **restoring force** in these systems. These systems are said to possess a **stable equilibrium** and the potential energy of these systems have the shape of a bowl as shown in Fig. 3.4.

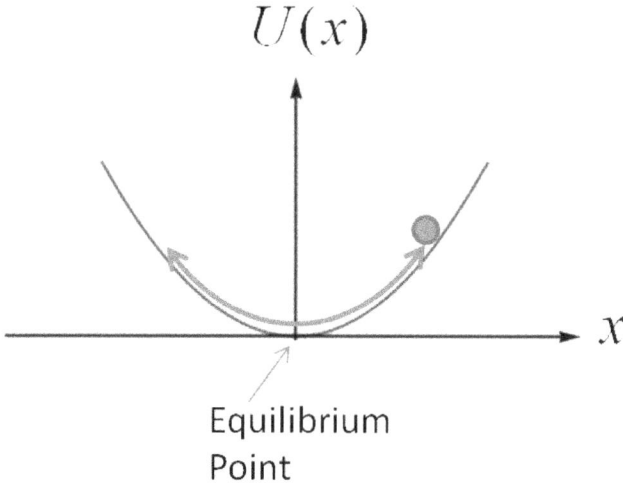

Fig. 3.4: Stable equilibrium occurs near minimum of the potential energy. The force corresponding to this potential energy will be pointed towards the left for points on the right side of the equilibrium and towards the right for points on the left side of the equilibrium.

When the system is in an equilibrium state, there is no net force on the system, but when the system is displaced from the equilibrium, a restoring force develops that acts on the system. The restoring force acts to bring the system back to the equilibrium state. However, when the system returns to the equilibrium point, it overshoots and goes past the equilibrium point due to the motional inertia, causing the restoring force to act again, but this time in the opposite direction. In this way, a back and forth or oscillatory motion takes place when a system is released at a point that is away from a stable equilibrium point.

3.1.2 Analytic Study

To study the oscillatory motion of the block attached to a support through a spring analytically, we choose a coordinate system such that the origin is located at the equilibrium position and the positive x-axis is pointed to the right as shown in Fig. 3.5. Let m be the mass of the block and k be the spring constant of the spring. For simplicity, we will ignore the mass of the spring in our treatment here.

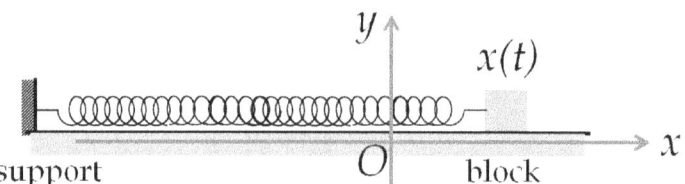

Fig. 3.5: The coordinate system for the study of oscillations of a mass attached to a spring with origin at a point on the table where the equilibrium point is located. The position of the block at an arbitrary time t is given by its x-coordinate at that time. The table is frictionless so that the only force on the block is the spring force. The y and z-components of net force are zero.

As mentioned above, the forces on the block in the vertical direction cancel out, and we are left with only the spring force on the block. At an arbitrary time t, the x-component of the net force on the mass is from the change in the length of the spring. The difference between the x-coordinates of the block and the support is the length of the spring. Let x_s be the x-coordinate of the spring. Then, the length l_0 of the spring when at equilibrium state and the length l of the spring at an arbitrary time are

$$\text{Relaxed state: } l_0 = 0 - x_s = -x_s$$
$$\text{Arbitrary state: } l = x - x_s$$

Therefore, the change in the length of the spring at arbitrary time is

$$\Delta l = l - l_0 = x.$$

Therefore, the magnitude of the spring force $\vec{F}_s$ will be

$$\boxed{\text{Magnitude of spring force} \quad |\vec{F}_s| = k\Delta l.}$$

The change in length of the spring can be written in terms of the x-coordinate of the block. Since, $x = 0$ corresponds to no change in the length of the spring, the absolute value of the x-coordinate of the block is equal to the change in the length of the spring.

$$\Delta l = |x|.$$

To write the x-component of the spring force in terms of the x-coordinate of the block, we notice that, when $x > 0$ the spring is stretched and the spring force is pointed towards the negative x-axis. When $x < 0$, the spring is compressed and the spring force is pointed towards the positive x-axis. Therefore, the x-component of the spring force will be given by multiplying the x-coordinate of the block by minus one.

$$\boxed{F_x = -k\ x.} \tag{3.1}$$

Newton's second law then gives us the following x-component of the equation of motion.

$$\boxed{m\ a_x = -k\ x,} \tag{3.2}$$

where a_x is the x-component of the acceleration of the block. We see that the acceleration of the block is proportional to the displacement of the block and points in the opposite direction to the position vector. Therefore, the acceleration of the block is not constant but depends on where the block is at the time.

A word of warning for students in order here. Since the acceleration of the block is not constant, you cannot use the formulas for the constant acceleration motion for a block attached to a spring. In particular,

$$\boxed{v_x \neq v_{0x} + a_x t}$$

$$\boxed{x - x_0 \neq v_{0x}t + \frac{1}{2}a_x t^2}$$

To understand and solve for the position of the block as a function of time, we find it helpful to write the acceleration in Eq. 3.2 as second derivative of position with respect to time. This makes the dependence of x with time more explicit.

$$\boxed{m \, \frac{d^2 x}{dt^2} = -k \, x.}$$ (3.3)

Thus, the equation of motion of the block is a second-order ordinary differential equation. The solution of this equation will give us the position of the block at a particular time. When we specify the initial position and velocity we can obtain a unique solution of Eq. 3.3. The solution, written as position as a function of time, $x(t)$, gives us the position of the block at all times. You will learn to solve this equation systematically in a more advanced course in mathematics, but in the next subsection we will solve this equation by guessing the answer and some reasonable arguments.

3.1.3 Solving Equation of Motion

To solve Eq. 3.2 or 3.3 means finding $x(t)$ and $v_x(t)$ when the position and velocity of the block are given at another time, say at $t = 0$. To carry out the calculations, it is helpful to combine the parameters m and k into one parameter.

$$\boxed{\frac{d^2 x}{dt^2} = -\omega^2 \, x,}$$ (3.4)

where we have introduced

$$\boxed{\omega^2 = k/m.}$$ (3.5)

The square for ω makes sure that k/m correspond to a positive number since both k and m are positive. We will see below that $|\omega|/2\pi$ is the frequency of the oscillator. We will often refer to ω itself as frequency, although it should be more appropriately called the **angular frequency**, since for each cycle ω changes by 2π radians. The frequency itself corresponds to the number of cycles per unit time and not to any radians per unit time. We will frequently ignore the problem with the naming and probably call ω frequency and hope that, from the context, you would understand that we are talking about the angular frequency.

Note that Eq. 3.4 shows that the displacement $x(t)$ from equilibrium is a function of time t, whose second derivative gives us back the same function multiplied by a negative constant. You may know from Calculus that there are two functions with this property: sine and cosine.

$$\frac{d^2 \cos(\omega t)}{dt^2} = -\omega^2 \cos(\omega t), \tag{3.6}$$

$$\frac{d^2 \sin(\omega t)}{dt^2} = -\omega^2 \sin(\omega t). \tag{3.7}$$

Any linear combination of sine and cosine with the same arguments will also work, as you can easily verify for the following function $f(t)$.

$$f(t) = A \, \cos(\omega t) + B \, \sin(\omega t)$$

$$\frac{d^2}{dt^2} f(t) = -\omega^2 \, f(t). \tag{3.8}$$

Thus, a general solution of Eq. 3.4 can be written as

$$\boxed{x(t) = C_1 \, \cos(\omega t) + C_2 \, \sin(\omega t),} \tag{3.9}$$

where C_1 and C_2 are arbitrary constants, which are determined from the initial position and initial velocity of the block. Different initial condition are different ways the oscillator can get started. The solution given in Eq. 3.9 can also be written in the following form.

$$\boxed{x(t) = A \, \cos(\omega t + \phi).} \tag{3.10}$$

where A and ϕ are constants. When the solution is written in the form of Eq. 3.10 or as $x = A \sin(\omega t + \phi)$, then the constant A is called the **amplitude** and ϕ the **phase constant**. The amplitude and phase constant depend on the initial conditions on the block just as C_1 and c_2. As a matter of fact by demanding that the two solutions give the same function $x(t)$, you can find relation between the two ways of writing the solution given in Eqs. 3.9 and 3.10 by deducing the relations between (C_1, C_2) pair and (A, ϕ) pair.

$$A = \sqrt{C_1^2 + C_2^2}, \quad \tan\phi = \frac{C_2}{C_1}. \tag{3.11}$$

You might say that C_1 and C_2 are "Cartesian" and A and ϕ are "polar" if you think of C_1 as some "x", C_2 as some "y", A as "r" and ϕ as "θ". We say that the pair (A, ϕ) is a "vector" of magnitude A and angle ϕ with respect to x-axis and C_1 and C_2 are the x and y-components of this abstract vector. The (A, ϕ) representation is also called a phasor representation.

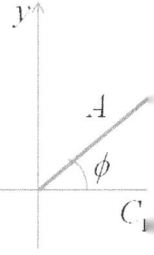

Definition of Simple Harmonic Motion

Equation 3.4 defines Simple Harmonic Oscillator. Any dynamical variable whose dynamics follows a differential equation that can me mapped on Eq. 3.4 is called a **Simple Harmonic Oscillator (SHO)**. As we have seen here, the dynamical variable can be written as a linear combination of sines and cosines of the same argument. The solution Eq. 3.9 or 3.10 is also said to describe a **Simple Har-**

monic Motion (SMH) because the solution implies a dynamics governed by Eq. 3.4.

3.1.4 Specifying Initial Position and Velocity

The initial position and velocity of the block determine the amplitude and phase constant of the motion as we show in this section. Let the initial position and velocity be

$$x(0) = x_0; \; v_x(0) = v_{0x}. \tag{3.12}$$

Working with solution Eq. 3.9: If we use the solution in the form given in Eq. 3.9 the initial condition can be used to find the constants C_1 and C_2 for the given oscillator. When we set $t = 0$, the sine term becomes zero.

$$x_0 = C_1. \tag{3.13}$$

Taking the time derivative turns cosine into sine and vice versa. Therefore the velocity at initial time is related to C_2.

$$v_{0x} = \omega C_2. \tag{3.14}$$

Finally, the solution incorporating initial conditions on position and velocity is

$$\boxed{x(t) = x_0 \, \cos(\omega t) + \frac{v_{0x}}{\omega} \, \sin(\omega t),} \tag{3.15}$$

Working with solution Eq. 3.10: Using the initial condition on position, we find the following from Eq. 3.10

$$x_0 = A \, \cos\phi. \tag{3.16}$$

To use the initial condition on velocity, we first take the first derivative of x with respect to time and then set $t = 0$.

$$v_{0x} = v_x(0) = \left. \frac{dx}{dt} \right|_{t=0} = -A \, \omega \, \sin\phi. \tag{3.17}$$

From Eqs. 3.16 and 3.17 we find the following for A and ϕ.

$$A = \sqrt{x_0^2 + (v_{0x}/\omega)^2} \; ; \; \tan\phi = -\frac{v_{0x}}{\omega x_0}. \tag{3.18}$$

3.1.5 Physical Meaning of Amplitude, A

The amplitude represents the maximum displacement of the oscillator from the equilibrium on either side of the equilibrium as a plot of the solution in Fig. 3.6 shows. The block turns around at $x = \pm A$, which are the **turning points** of motion. The velocities of the block at the turning points are zero.

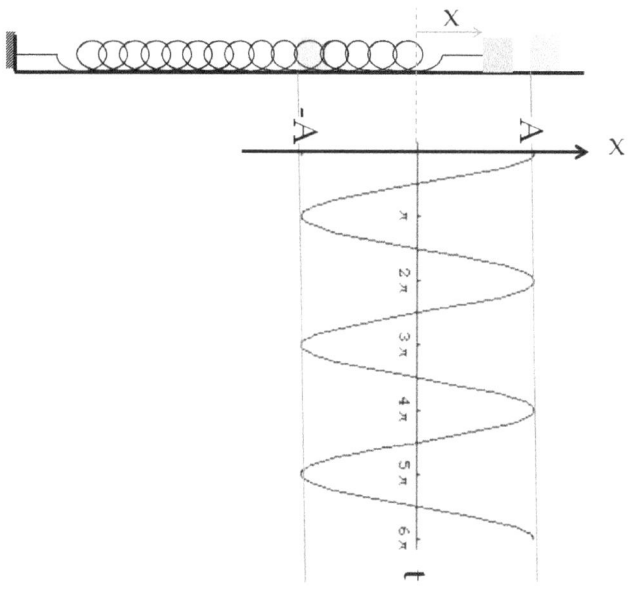

Fig. 3.6: Plot of x vs t for $\omega = 1$, and $\phi = 0$. A one-dimensional harmonic oscillator moves within a displacement A from the equilibrium on either side.

When the block approaches the turning point on the right its velocity is positive and decreasing since the force on the block would be pointed to the left and increasing. At the turning point, the force, and hence the acceleration of the block, is the largest and the velocity is zero. After the block's motion turns around at the turning point $x = A$, the velocity and acceleration both point to the left, which accelerates the block to the equilibrium point at $x = 0$. There is a similar turning of the motion near the other turning point at $x = -A$.

3.1.6 Physical Meaning of ω

The solution of simple harmonic motion is periodic in time since it describes the periodic motion of the mass attached to the spring. Let T be the time for one complete cycle of motion. The time T is called the **time period** of the oscillator. We expect both the position and the velocity to be periodic functions of time with the same time period T.

$$x(t + T) = x(t) \quad \text{and} \quad v_x(t + T) = v_x(t). \tag{3.19}$$

Demanding the periodicity in $x(t)$ given in Eq. 3.10 we find that ω is related to the period:

$$A\cos(\omega t + \omega T + \phi) = A\cos(\omega t + \phi).$$
$$\implies \omega T = 2n\pi, \quad n = 0, \pm 1, \pm 2, \cdots, \tag{3.20}$$

Clearly $n = 0$ is not the solution since that would mean T be zero, which is not the case here. The $n = 1$ case refers to a situation after one time period, and $n = -1$, to a situation one time period earlier. The constant ω is therefore 2π over the time period.

The inverse of time period gives the number of cycles of oscillations completed by the oscillator in a unit time. This is called the **frequency** of the oscillator. We will denote the frequency by the letter f.

$$f = \frac{1}{T}. \tag{3.21}$$

The SI unit of frequency is 1/sec, which is given its own name, **the Hertz (Hz)**. Therefore, ω can be written in terms of frequency f as

$$\omega = 2\pi f. \tag{3.22}$$

This says that ω "counts" radians in unit time instead of cycles in unit time. Therefore, we call ω an angular frequency and f regular frequency. Because f is number of cycles per unit time, f is a positive number. Since ω is a positive number times f, ω will also be positive. Therefore when we take square-root of ω^2, we keep only the positive root.

$$\omega = 2\pi f = \sqrt{\frac{k}{m}}. \tag{3.23}$$

The angular frequency ω depends only on the mass and the spring constant. That is why ω is also called the **natural frequency** of the oscillator.

3.1.7 Analogy to Circular Motion

Some aspects of a particle moving in a circle provide useful tools for visualizing the Simple Harmonic Motion in another way. If a particle is moving in a circle of radius r about the origin and be at an angle θ at time t, then the x and y-coordinates of the position would be given by $r\cos(\theta)$ and $r\sin(\theta)$ respectively.

We found above that the displacement of a Simple Harmonic Motion can be given by a cosine function of time. Therefore, we can represent a simple harmonic motion by the x-component of the motion of a fictitious particle moving uniformly in a circle as shown in Fig. 3.7. In this picture of a Simple Harmonic Motion, ω refers to the angular speed of the fictitious particle. Since the fictitious particle moves uniformly, the magnitude of its angular displacement $\Delta\theta$ of the fictitious particle in time interval Δt would be given by

$$\Delta\theta = \omega\Delta t. \tag{3.24}$$

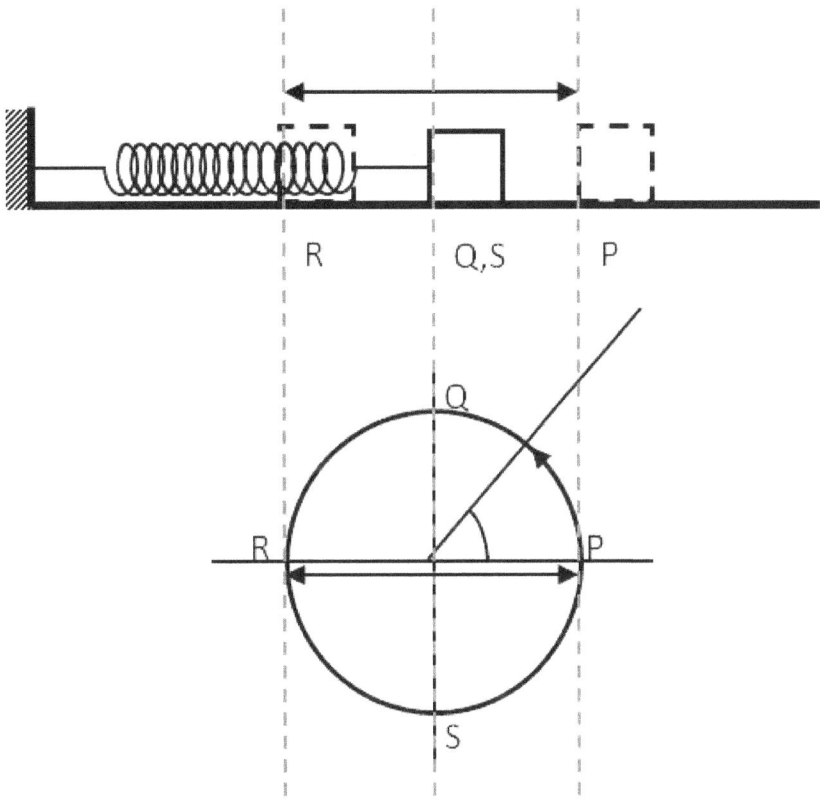

Fig. 3.7: The cyclic process of oscillatory motion of the block represented as a point moving on a circle. The angular frequency ω corresponds to the angle in radians covered by the point on the circle per unit time, so that angle covered in time Δt is $\omega\Delta t$. Here P, Q, R, S are successive positions of the block as it executes simple harmonic motion.

3.1.8 Physical Meaning of Phase Constant, ϕ

The phase constant ϕ is related to the relative position of the block in its cycle if the cycle is represented by a rotation of 2π radians of a fictitious particle in the circular motion as explained above. To get a feel for the physical information contained in the phase constant, we examine two oscillators with $\phi_1 = 0$ and $\phi_2 = \frac{\pi}{2}$ radians respectively as shown in Fig. 3.8. Notice that the second oscillator is always ahead of the first by quarter of a cycle: for instance, $x = A$ is reached by oscillator 2 before oscillator 1 in any cycle. In terms of 2π radians in one cycle, this corresponds to a phase difference of $\frac{\pi}{2}$ radians.

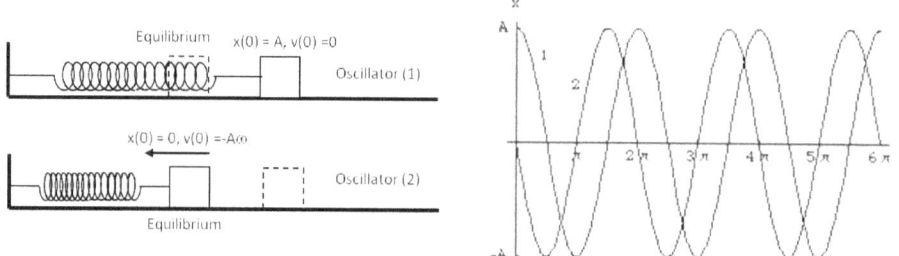

Fig. 3.8: Oscillators started in two different ways such that they have the same amplitude but have different phase constants. We plot their position in time for $\omega = 1$ rad/sec and the following initial conditions: (1) $x = A, v = 0$, and (2) $x = 0, v = -A$. The relative phase different is $\frac{\pi}{2}$ radians or $\frac{1}{4}$ cycle.

Thus, the phase constant ϕ represents a measure of time in the cycle of the oscillator as measured in terms of angle of the fictitious particle moving in a circle. You may practice drawing displacements of two oscillators of same frequency and amplitude but differing in phase by π radians or $\frac{\pi}{3}$ radians and see for yourself the relation between oscillators.

3.1.9 Energy of a Simple Harmonic Oscillator

The energy of the block is the sum of the kinetic and potential energies of the block. The kinetic energy of the block is given as

$$KE = \frac{1}{2}mv^2, \tag{3.25}$$

and the potential energy of the block for the spring force is given as

$$PE = \frac{1}{2}kx^2, \tag{3.26}$$

with the reference for the potential energy set at zero when the spring is relaxed. Now, since the block oscillates sinusoidally, meaning as a sine or cosine function, the kinetic and potential energies oscillate sinusoidally also. However, the total energy E of the oscillator is constant since the spring force is a conservative force.

$$E = KE + PE = \frac{1}{2}mv^2 + \frac{1}{2}kx^2 = \text{constant.} \qquad (3.27)$$

You can verify that $\frac{1}{2}mv^2 + \frac{1}{2}kx^2$ is constant for a harmonic oscillator by using either Eq. 3.9 or 3.10 for $x(t)$.

Equation 3.27 asserts that the energy of a Simple Harmonic Oscillator contains two terms, an inertial term that is quadratic in velocity and an elastic restoring term that is quadratic in displacement. Any system whose conserved energy can be written this way can be treated as a Simple Harmonic Oscillator (SHO). Suppose the dynamical variable of an abstract system is $f(t)$ and its time derivative is $\dot{f}$, then, this system will be an SHO if its energy takes the following mathematical form for some constant a and b.

$$E = a\dot{f}^2 + bf^2. \qquad (3.28)$$

From the expression for the energy of oscillator given in Eq. 3.27, it is clear that the expression for the energy can also be used to deduce the frequency of the oscillator. The ratio of the constant $(k/2)$ multiplying the displacement squared and the constant $(m/2)$ multiplying the velocity squared is equal to the angular frequency squared.

$$\omega^2 = \frac{\text{Coefficient of } x^2}{\text{Coefficient of } v^2} = \frac{k}{m}. \qquad (3.29)$$

As the block gets closer to the extreme points of the motion, the magnitude $|x|$ of the displacement from equilibrium becomes closer to the maximum. At the turning points, the potential energy has the largest value, since at these points, there is no kinetic energy and all the energy must be in the potential energy. When the block returns to the point which corresponds to a point when the spring is relaxed, then the potential energy of the block will be zero and the energy would all be in the kinetic energy.

$$E = \frac{1}{2}mv^2 + \frac{1}{2}kx^2 = \frac{1}{2}kA^2 = \frac{1}{2}mv_{\text{max}}^2, \qquad (3.30)$$

where v_{max} is the speed when the block passes the equilibrium point. This energy conservation is for a block attached to a spring and placed on a frictionless table and the other end of the spring is attached to a rigid support.

Note that if there are other forces on the block in the block/spring system, then the zero potential energy and zero force may not be the same place. We will analyze the hanging mass attached to a spring later in the chapter where there will be two forces along the axis of the spring, the spring force and gravity.

Example 3.1. Conservation of Energy of a Harmonic Oscillator. *A block of mass 0.6 kg is attached to a spring of spring constant 180 N/m and negligible mass compared to the mass of the block. The block is placed on a frictionless horizontal table, pulled horizontally 3 cm from its equilibrium position, and released with zero speed. Evaluate the speed of the block at the time (a) the block crosses the equilibrium position, and (b) when the block is 1.5 cm from the equilibrium point.*

Solution. Since no friction acts on the block, the energy is conserved. Let us denote the original position as point A, the equilibrium as point B and 1.5 *cm* from the equilibrium as point C.

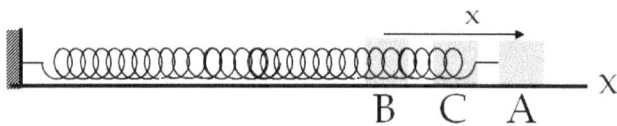

Fig. 3.9: Example 3.1

(a) When the block is at point A, it is not moving, therefore all the energy is in the potential energy stored in the spring.

$$E_A = \frac{1}{2}kx_A^2 = \frac{1}{2}(180 \text{ N/m})(0.03 \text{ m})^2 = 0.081 \text{ J}.$$

When the block is at point B, the spring is neither stretched not compressed, therefore there is no potential energy and the entire energy is contained in the Kinetic energy.

$$E_B = \frac{1}{2}mv_B^2 = \frac{1}{2}(0.6 \text{ kg})v_B^2.$$

Equating the energy at B to the energy at A we find the speed of the block when it is moving past the equilibrium point B.

$$0.3 \ v_B^2 = 0.081 \Rightarrow v_B = 0.52 \text{ m/s}.$$

(b) When the block is at point C, which could be on either side of the equilibrium a distance of 1.5 cm away from the equilibrium, the spring is either compressed or stretched. Therefore, there will be potential energy stored in the spring. But, the stored potential energy is less than the starting energy, hence, the rest of the energy will be in the form of kinetic energy of the block.

$$E_C = \frac{1}{2}mv_C^2 + \frac{1}{2}kx_C^2 = 0.3 \ v_C^2 + 0.02 \text{J}.$$

Equating E_C to E_A, and solving for v_C we find the speed when the block is 1.5 cm from the equilibrium to be

$$0.3\, v_C^2 + 0.02 = 0.081 \Rightarrow v_C = 0.45 \text{ m/s}.$$

3.1.10 Vertically Oscillating Block

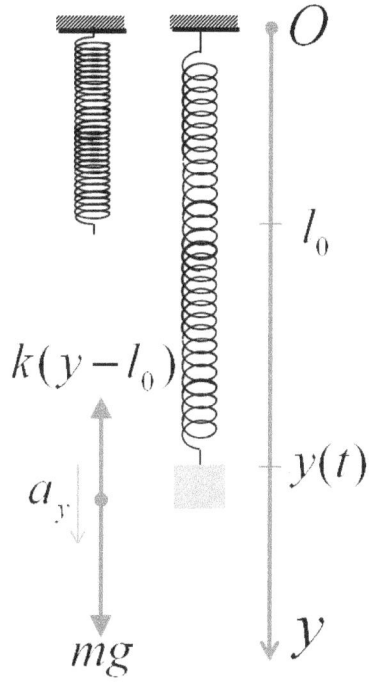

Fig. 3.10: Choice of origin and axis for the calculations regarding a vertically oscillating block attached to a spring.

We now examine a block attached to a spring but hung vertically by fixing the other end to a support in the ceiling. There are two forces on block in the vertical direction and none in the horizontal direction. Let m be the mass of the block and k the spring constant of the spring. We also assume that the mass of the spring is negligible. Will the block have the same frequency of oscillation and same energy formulas as the block on the horizontal frictionless table discussed above?

To analyze the motion of the block we need to choose the origin and one axis. We will choose a coordinate axes so that the positive y-axis is pointed vertically down as shown in Fig. 3.10. Pointing y-axis vertically down will be helpful in writing out the expression for the spring force as you will see below.

Where should we place the origin? Note the force by the spring is proportional to the change in the length of the spring with respect to the length the spring has when it is relaxed. Suppose we place the origin at the ceiling where the spring is held. This is a convenient place for out initial calculations as you will see below. We may end up choosing another place for the origin after we have understood the calculations a little bit more.

Let l_0 be the relaxed length of the spring when it is hanging from the ceiling and the block is not attached. When the block is attached, the spring stretches. Let y be the position of the block at an arbitrary time t. Then, the change in the length of the spring at this time is given by $y - l_0$. The y-component F_y of the spring force is now readily written down.

$$F_y = -k(y - l_0).$$

Let us make sure this gives correct directions for the force. When $y - l_0 > 0$, then the spring is stretched and so the force should be pointed up. That means towards negative y-axis. That works out with the negative sign in this formula. You should do a similar analysis to make sure the formula works for $y - l_0 < 0$.

The free-body diagram of the block with acceleration given by two time derivatives of the y-coordinate gives the equation of motion along y-axis as

$$m\frac{d^2y}{dt^2} = -k(y - l_0) + mg. \qquad (3.31)$$

Note the y-component of weight is positive since y-axis is pointed down. This equation is more complicated than the equation for the block on the horizontal table. This equation is simpler if written with respect to origin at the equilibrium point of the block. We can find the y-coordinate of the equilibrium point by setting the acceleration in Eq. 3.31 to zero. Let the equilibrium be at $y = y_e$. Then we find that

$$-k(y_e - l_0) + mg = 0,$$

which gives

$$y_e = l_0 + \frac{mg}{k}. \qquad (3.32)$$

To move the origin to the new place, all we have to do is change to new y-variable, say y' which is related to the old y by

$$y' = y - y_e. \qquad (3.33)$$

Therefore, y' will be the y-coordinate from the point in space where the equilibrium is located. The acceleration in the y' coordinate is equal to the acceleration in the y-coordinate as you can see by taking two time derivatives in this equation. Now, let us substitute Eq. 3.33 into Eq. 3.31. We find

$$m\frac{d^2y'}{dt^2} = -ky' \qquad (3.34)$$

The equation with respect to the equilibrium is same as the Simple Harmonic Oscillator (SHO) equation. Therefore, we will have the same solution for the y' coordinate as for any other SHO.

$$y'(t) = A\cos(\omega t + \phi) \quad \text{or} \quad C_1\cos(\omega t) + C_2\sin(\omega t), \qquad (3.35)$$

with angular frequency, $\omega = \sqrt{k/m}$ as obtained from Eq. 3.34.

Energy Considerations in Vertical Oscillations: The energy of the block will again be equal to the sum of the kinetic and potential energies of the block. Here, there are two sources of potential energy. Therefore, the energy can be written as

$$E = \frac{1}{2}mv(t)^2 + \frac{1}{2}k\left[\Delta l(t)\right]^2 + mgh(t) \qquad (3.36)$$

where $\Delta l(t)$ is the change in the length of the spring at time t, $h(t)$ is the height of the block at that instant from a reference height for gravitational potential energy and $v(t)$ is the speed of the block at that time. We can write more explicit expressions of the potential energy with respect to the origin at the ceiling as the reference for the zero of the gravitational energy. With this choice, the gravitational potential energy is $-mgy$, where y is the y-coordinate of the block. The change in the length of the spring can be written as $y - l_0$ in this coordinate system. Therefore, the energy of the block in the coordinate system with origin at the ceiling and y-axis pointed down is

$$E = \frac{1}{2}m\left(\frac{dy}{dt}\right)^2 + \frac{1}{2}k\left[y(t) - l_0\right]^2 - mgy(t). \qquad (3.37)$$

This can be written in terms of y' coordinate of the block as

$$E = \frac{1}{2}m\left(\frac{dy'}{dt}\right)^2 + \frac{1}{2}ky'^2 - \frac{mg}{2k}\left(mg + 2kl_0\right), \qquad (3.38)$$

which shows that the potential energy in the y' does not the simple $1/2kx^2$ type formula in the case of vertical oscillations. We can make another change in variable to "discover" the SHO nature of this system in the energy picture as well. Let

$$u(t) = y' - \frac{1}{k}\sqrt{mg(mg + 2kl_0)}.$$

In the u variable which shifts the origin to yet another place on the y-axis, the energy becomes

$$E = \frac{1}{2}m\left(\frac{du}{dt}\right)^2 + \frac{1}{2}ku^2.$$

3.2 EXAMPLES OF SIMPLE HARMONIC MOTION

Simple harmonic motion appears in a wide variety of physical settings. When a system is displaced from a stable equilibrium, the restoring force is usually proportional to the displacement if the displacement is not too great. In this section, we study some commonly encountered systems that exhibit simple harmonic motion.

Plane Pendulum

A pendulum consists of a bob of mass m suspended from a light inextensible cord of length l. If the physical dimension of the bob is much smaller than the length of the cord, we can treat the bob as a point mass. A pendulum has a stable equilibrium position when the bob is hanging vertically down from the suspension point. The angular displacement of the bob from the equilibrium is given in terms of angle θ that the cord makes with the vertical line as shown in Fig. 3.11. The displacement angle is positive for the counter-clockwise change in angle and negative for a clockwise change in angle when viewed from the axis coming out-of-page in the figure.

Equation of Motion Approach:

Because we wish to study the angular displacement, it is more convenient to treat the pendulum problem as a rotation problem — the rotation of the point mass m about an axis through the point of suspension O and perpendicular to the plane of oscillation. We choose a Cartesian coordinate system as shown in Fig. 3.11 so that angle θ corresponds to the z-component of the angular displacement, θ_z. The z-component of the equation of motion for rotation based on the z-components of the torque and angular acceleration, and the I_{zz} component of the moment of inertia.

$$\tau_z = I_{zz}\frac{d^2\theta}{dt^2}. \tag{3.39}$$

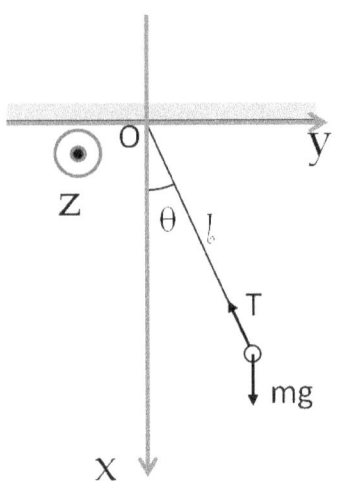

Fig. 3.11: A plane pendulum oscillates in a plane, rotating about an axis through the point of suspension O and perpendicular to the plane of the pendulum motion. The angle θ that the cord makes with the vertical line gives the displacement of the pendulum from a stable equilibrium.

To find the torque on the bob, we note that there are only two forces on the bob: the weight of the bob and the tension in the cord. Since the force of tension goes through the point of suspension O, it does not exert any torque about O. The torque about O comes only from the weight, which gives the following z-component:

$$\tau_z = -mgl\sin\theta, \tag{3.40}$$

where the minus sign refers to the fact that torque is pointed towards the negative z-axis when θ is positive and towards the positive z-axis when θ is negative. The moment of inertia component I_{zz} of the bob about the axis is

$$I_{zz} = m(x^2 + y^2) = ml^2. \tag{3.41}$$

Therefore, the z-component of the rotational equation of motion of a pendulum is

$$\frac{d^2\theta}{dt^2} = -\frac{g}{l}\sin\theta. \tag{3.42}$$

Here, the angular acceleration is not proportional to the angular displacement, but to the sine of the angular displacement. Therefore, a plane pendulum does not execute a Simple Harmonic Motion, which requires that the acceleration be proportional to displacement. Another problem with the equation of motion of pendulum, Eq. 3.42, is that it is not easily solvable. But, we note that for small angles (expressed in radians), the sine of the angle can be approximated by the angle itself.

$$\sin(\theta) \approx \theta \text{ (small angles in radians).} \qquad (3.43)$$

With the small angle approximation, the equation of motion of a pendulum, Eq. 3.42, can be simplified to

$$\frac{d^2\theta}{dt^2} = -\frac{g}{l}\theta, \qquad (3.44)$$

where I have replaced the approximate sign ($\approx$) with the equal sign ($=$) since we intend to be within the limit of the applicability of the approximation being utilized. The modified equation of motion, Eq. 3.44 shows clearly that, if the oscillations are kept to small angles, the angular acceleration (two-time derivatives of θ) is proportional to the angular displacement (θ) and is opposed to the angular displacement. Therefore, for small-angle displacements, a pendulum will execute a simple harmonic motion.

Comparing the approximate equation of motion for a pendulum with the equation for a block attached to a spring, we immediately discover the following correspondences in symbols.

$$x \Longleftrightarrow \theta \qquad (3.45)$$
$$(k/m) \Longleftrightarrow (g/l) \qquad (3.46)$$

Exploiting this analogy, we can write the solution of the small angle pendulum problem as

$$\theta(t) = A \cos(\omega t + \phi) = C_1 \cos \omega t + C_2 \sin \omega t, \qquad (3.47)$$

and deduce the formula for the angular frequency of the plane pendulum:

$$\omega = \sqrt{\frac{g}{l}}, \qquad (3.48)$$

which gives the period of the pendulum to be:

$$T = \frac{2\pi}{\omega} = 2\pi \sqrt{\frac{l}{g}}. \qquad (3.49)$$

The formula for the period shows that the time period of a pendulum does not depend on either the mass m of the pendulum bob or the amplitude A of oscillations, but only on the length l of the pendulum cord and acceleration due to gravity g. The dependence of the period on g means that the same pendulum will have different periods on different planets. Also, since the value of g varies over the surface of the Earth, the same pendulum will run at different rate in different locations on the Earth.

Galileo appears to be the first person who noticed this aspect of a pendulum motion when he made the observation that different

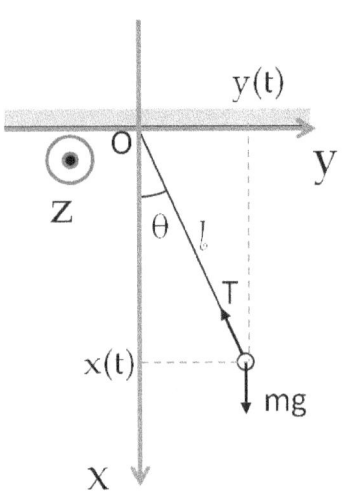

Fig. 3.12: The x and y motions of the plane pendulum.

chandeliers of equal length in a church had the same period regardless of their amplitudes of swing or weights. Apparently Galileo timed the swings of chandeliers using his pulses and found that chandeliers swung at the same period regardless of their masses and swings.

Energy Approach:

The formula for the period of the small oscillations of the pendulum can be more easily deduced by examining the energy of the bob as we will show next. The energy of a pendulum can also be written such that it contains two terms, one having the square of velocity for the kinetic energy, and the other the square of displacement for the potential energy, similar to that of a Simple Harmonic Oscillator. Let x and y denote the x and y-coordinates of the pendulum bob at an arbitrary time t with respect to the axes shown in Fig. 3.12. Then, in the small angle approximation, we have

$$x = l - l\,\cos\theta \approx \frac{l}{2}\theta^2, \qquad (3.50)$$

$$y = l\,\sin\theta \approx l\theta, \qquad (3.51)$$

where we have dropped terms which are cubic or higher powers in θ. The velocities are

$$v_x = \frac{dx}{dt} = l\theta\,\frac{d\theta}{dt}, \qquad (3.52)$$

$$v_y = \frac{dy}{dt} = l\,\frac{d\theta}{dt}. \qquad (3.53)$$

The kinetic energy of the bob is found to be

$$KE = \frac{1}{2}m\left(v_x^2 + v_y^2\right) \approx \frac{1}{2}ml^2\left(\frac{d\theta}{dt}\right)^2, \qquad (3.54)$$

where we have dropped the term with four powers involving θ and $d\theta/dt$. The potential energy is

$$PE = mg(l - x) = mg\frac{l}{2}\theta^2. \quad \text{(reference zero at the lowest point)} \qquad (3.55)$$

Therefore, the energy of the pendulum in small angle approximation is

$$E = \frac{1}{2}ml^2\left(\frac{d\theta}{dt}\right)^2 + \frac{1}{2}mgl\theta^2, \qquad (3.56)$$

which has the required form for a simple harmonic motion. Hence, we can also determine the angular frequency also from the ratio of the coefficients of the restoring and inertial terms.

$$\omega = \sqrt{\frac{mgl}{ml^2}} = \sqrt{\frac{g}{l}}, \qquad (3.57)$$

which is identical to the expression we obtained from an application of the second law of motion as given in Eq. 3.48.

Torsion Pendulum

A torsion pendulum consists of a solid, a dumbbell, a disk, a bar, or an object of any other shape, suspended by a torsion wire from a fixed support as shown in Fig. 3.13. A torsion wire is essentially a wire that could be twisted easily about its length. The twisting of the wire applies a restoring torque on the supported body whose tendency is to bring the body back to the configuration when the wire was not twisted, i.e., to the equilibrium. According to the Hooke's law for elasticity of materials, the restoring torque τ on the bar should be proportional to the angle of twist, at least for the small angles of twist we would work with. Using a coordinate system in which the z-axis is pointed in the direction of the axis of rotation, the z-component of the torque would be

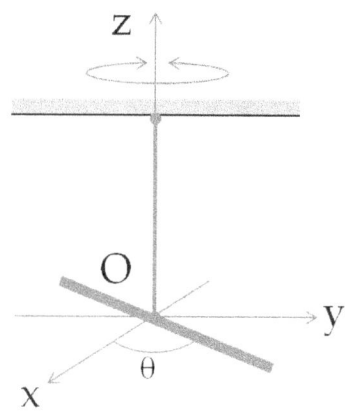

$$\tau_z = -\kappa\theta, \tag{3.58}$$

Fig. 3.13: The torsion pendulum.

where κ (read: kappa) is the torsional constant of the wire, and θ is the z-component of the angular displacement from equilibrium as measured from x-axis in the xy-plane. The torsional constant for twisting wire is analogous to the spring constant of a spring. The rotational motion of the bar is then given by the z-component of the rotational equation of motion.

$$I_{zz}\frac{d^2\theta}{dt^2} = -\kappa\theta, \tag{3.59}$$

where I_{zz} is the moment of inertia of the bar about z-axis, which is the axis of rotation. This equation is analogous to the equation for a plane pendulum in small angle approximation. By exploring the mathematical correspondence of symbols in the equations here to the equations of the simple pendulum we can read off the expression for the angular frequency ω of oscillating motion of the torsion pendulum as

$$\omega = \sqrt{\frac{\kappa}{I_{zz}}}. \tag{3.60}$$

Torsion pendulums are often used for time-keeping purposes, e.g., in the balance wheel of a mechanical watch. A torsion pendulum is also used in Cavendish experiment for determining the value of the Newton's gravitational constant G_N.

Physical pendulum

A rigid body hung from a post swings just like a pendulum. Such oscillating bodies are called physical pendulums. Almost anything can be a physical pendulum. An illustration in shown in Fig. 3.14. The torque responsible for the oscillations comes from the force of gravity on the body, which is calculated by placing the weight vector at the center of mass of the body. What is the frequency of small oscillations in this case?

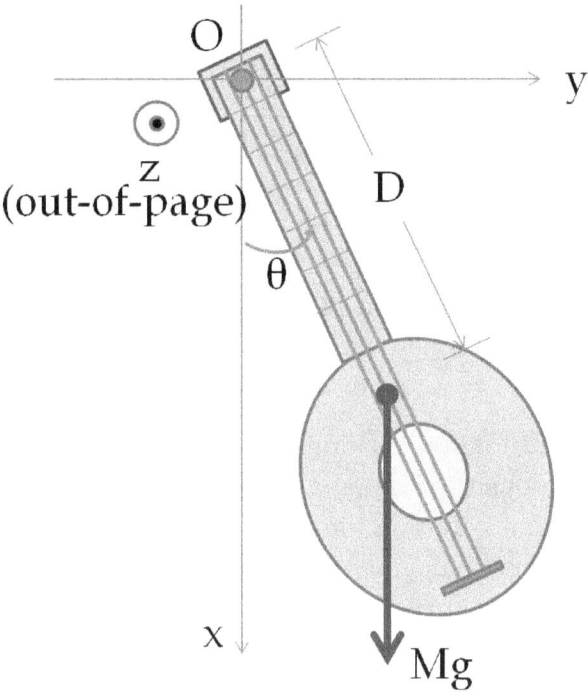

Fig. 3.14: A physical pendulum can be anything that can swing about an axis. Here a musical instrument oscillated about a fixed point at one end illustrates a physical pendulum.

Let M be the mass and D the distance between the suspension point and the CM of the physical pendulum. From the z-component of the torque due to gravity we obtain the following equation of motion for the rotation about the z-axis in terms of the angle θ with respect to the positive x-axis.

$$I_{zz}\frac{d^2\theta}{dt^2} = -MgD\sin\theta, \qquad (3.61)$$

where I_{zz} is the moment of inertia about z-axis passing through the point of suspension. The mathematical situation here is similar to that of the plane pendulum. Once again, we resort to the small angle approximation. For small-angle oscillations, we set $\sin\theta \approx \theta$ in Eq.

3.61, which yields the following approximate equation of motion of a physical pendulum.

$$\frac{d^2\theta}{dt^2} \approx -\frac{MgD}{I_{zz}}\theta.$$ (3.62)

Now, by analogy with the plane pendulum, we find that the angular frequency of oscillation of a physical pendulum is given by

$$\omega = \sqrt{\frac{MgD}{I_{zz}}}.$$ (3.63)

3.3 MOTION NEAR POTENTIAL MINIMA

From our discussion in this chapter, you know that that a restoring force that is proportional to the displacement from the equilibrium and points in the opposite direction will lead to a Simple Harmonic Motion. Now, the x-component of a conservative force $\vec{F}$ is related to potential energy U as follows,

$$F_x = -\frac{dU}{dx}.$$ (3.64)

Therefore, any potential energy that is quadratic in x, the displacement variable, will result in the restoring force appropriate for a Simple Harmonic Motion. This is obviously the case with the potential energy due to force from an ideal spring. In general, consider a potential energy $U(x)$ that has a minimum at $x = x_0$. By writing the potential energy function in terms of a Taylor series about $x = x_0$ we obtain the following.

For, a general case, we need to use partial derivatives.

$$\vec{F} = -\left(\frac{\partial U}{\partial x}\hat{u}_x + \frac{\partial U}{\partial y}\hat{u}_y + \frac{\partial U}{\partial z}\hat{u}_z\right),$$

where $\hat{u}_x, \hat{u}_y$, and $\hat{u}_z$ are unit vectors pointed towards the positive x, y and z-axes respectively.

$$U(x) = U(x_0) + \left(\frac{dU}{dx}\right)_{x=x_0}(x - x_0) + \frac{1}{2!}\left(\frac{d^2U}{dx^2}\right)_{x=x_0}(x - x_0)^2 + \cdots$$ (3.65)

Since the potential energy has a minimum at $x = x_0$, the first derivative is zero there, and the leading non-constant term is the quadratic term in $x - x_0$, the displacement from the equilibrium.

$$U(x) = U(x_0) + \frac{1}{2!}\left(\frac{d^2U}{dx^2}\right)_{x=x_0}(x - x_0)^2 + \cdots$$ (3.66)

The value of the second derivative of the potential energy function for $x = x_0$ is a constant. Let us denote this constant by k in anticipation of its analogy with the spring constant of a spring.

$$k \equiv \left(\frac{d^2U}{dx^2}\right)_{x=x_0}.$$ (3.67)

Choosing the potential energy to be zero at the equilibrium, and placing the origin at the equilibrium point, we find that near a potential energy minimum, the leading behavior of the potential energy function is quadratic.

$$U(x) = \frac{1}{2!}kx^2 + \cdots \tag{3.68}$$

Therefore, even though an oscillating system may not be a block attached to a spring, the behavior is "identical" to the problem of a block attached to a spring and we can speak of a "spring constant" whenever a system is oscillating such that near the bottom of the potential energy the potential energy can be approximated by a quadratic function of the corresponding displacement. The only exceptions are those potential energy functions, such as $U(x) = bx^4$, which cannot be approximated by a quadratic function near the minima.

A quadratic potential energy function gives a linear restoring force of the Hooke's law and leads to the Simple Harmonic Motion.

$$F_x = -\frac{dU}{dx} = -kx + \text{ higher powers in } x. \tag{3.69}$$

We have seen above that the plane pendulum is not a Simple Harmonic Oscillator unless the angle of oscillation is small. We can see this emerging Simple Harmonic property from the perspective of a quadratic potential energy function. The potential energy of a pendulum when it is displaced at angle θ is

$$U = mgl(1 - \cos\theta). \tag{3.70}$$

Now, expanding $\cos(\theta)$ for small θ we find

$$\cos(\theta) = 1 - \frac{1}{2!}\theta^2 + \frac{1}{4!}\theta^4 + \dots. \tag{3.71}$$

If we keep only the leading term, viz., 1, we will lose all physics information associated with θ. Therefore, we will keep two terms in this expansion. This gives the following expression for the potential energy near $\theta = 0$:

$$U = \frac{mgl}{2}\theta^2, \tag{3.72}$$

which is quadratic in the dynamical variable θ. Hence, for small angles we expect a Simple Harmonic Motion for the pendulum as discussed previously.

3.4 THE DAMPED HARMONIC OSCILLATOR

3.4.1 Under, over and critical damping

Frictional forces damp the motion of a moving object by taking energy away from the object. A particularly simple type of a viscous force whose magnitude is proportional to the velocity of the oscillator has many applications in physics of oscillatory systems.

$$\vec{F}_{visc} = -b\,\vec{v} \quad (b \geq 0), \tag{3.73}$$

where b is the proportionality constant, called the **viscous damping coefficient**, that depends on the viscous medium and the geometry of the oscillating mass. The minus sign makes sure that the viscous force is pointed in the opposite direction to the velocity. Thus, this force will slow the motion of the object. Viscous forces of this type act on objects moving in a fluid if their speeds are not too great. Pictorially, it is customary to represent the damping force by attaching a dash pot to the oscillating block attached to the spring as shown in Fig. 3.15.

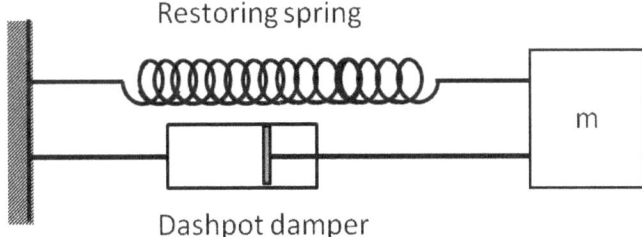

Fig. 3.15: A harmonic oscillator with a dashpot. The dashpot has a piston that moves through a liquid providing velocity dependent damping force.

The net force on the oscillator will now be the vector sum of the restoring force by the spring and the viscous force. To cast the problem in analytic terms, we choose Cartesian coordinates. Once again, we will place the origin at the equilibrium position of the block, and point the positive x-axis in the direction that corresponds to the extension of the spring. Then the x-component of the second law of motion would take the following form.

$$m\frac{d^2x}{dt^2} = -kx - b\,\frac{dx}{dt}, \tag{3.74}$$

where I have replaced the x-component of the acceleration by the second derivative of the x-coordinate of the block and the x-component of the velocity by the first derivative. It is helpful to divide both

Note that some authors use $\gamma = b/m$ in place of $\beta = b/2m$.

sides of this equation by m, and introduce new constants, β and ω_0 as follows so that the solution of Eq. 3.74 can be written with fewer parameters.

$$2\beta = b/m, \qquad (3.75)$$

$$\omega_0 = \sqrt{k/m}, \qquad (3.76)$$

where a factor of 2 in the definition of β is included to simplify the formulas that appear below. The quantity ω_0 is called the "natural frequency" of the oscillator. It refers to the innate oscillation frequency at which the block will oscillate if there were no damping or other forces on the block. The quantity β is called **damping constant** or damping parameter.

Note that now we are using a subscript zero for the natural frequency of the undamped oscillator because we will be encountering other frequencies in the same system, and we do not want to get confused about them. We will see below that a damped oscillator does not oscillate at the natural frequency, but with a new frequency depending on both β and ω_0.

With these new parameters, Eq. 3.74 can be written as

$$\frac{d^2x}{dt^2} + 2\beta\,\frac{dx}{dt} + \omega_0^2 x = 0. \qquad (3.77)$$

The oscillating characteristics of a damped oscillator depends on whether ω_0 is less than, equal to, or greater than β. If $\omega_0 > \beta$, then the mass oscillates about the equilibrium, successively damping out each cycle, and the system is said to be under damped; if $\omega_0 < \beta$, then the system does not oscillate at all, and we say that the system is overdamped; finally, if $\omega_0 = \beta$, the system is called critically damped, which separated the underdamped from the over-damped cases, and where, again, there is no oscillation. The mathematical expressions of the three solutions are as follows.

$$x(t) = \begin{cases} A_1 \exp\left(-\beta t\right)\cos(\omega_1 t + \phi) \text{ for } \omega_0 > \beta, \text{ under-damped} \\ \exp\left(-\beta t\right)(A_2 t + A_3) \text{ for } \omega_0 = \beta, \text{ critically damped} \\ \exp\left(-\beta t\right)\left[A_4 \exp\left(-\alpha t\right) + A_5 \exp\left(\alpha t\right)\right], \\ \qquad \text{for } \omega_0 < \beta, \text{ over-damped} \end{cases}$$

$$(3.78)$$

where

$$\omega_1 = \sqrt{\omega_0^2 - \beta^2} \quad \text{and} \quad \alpha = \sqrt{\beta^2 - \omega_0^2}. \qquad (3.79)$$

Constants A_1, A_2, A_3, A_4, A_5, and ϕ in Eq. 3.78 are determined by the initial position and initial velocity, as discussed for the undamped oscillator above. It would seem that if you want the oscillations to die out quickly, then may be, you should try overdamping the system.

It turns out that overdamping is not the right strategy. Instead, we find that an oscillator is damped more quickly if critically damped as shown in Fig. 3.17. In practical uses of the damping, for example in the shock absorbers for cars, you would want the car not to carry on with bouncing up and down whenever the hits a bump in the road. Therefore, one builds the shock absorbers with the characteristics of critical damping.

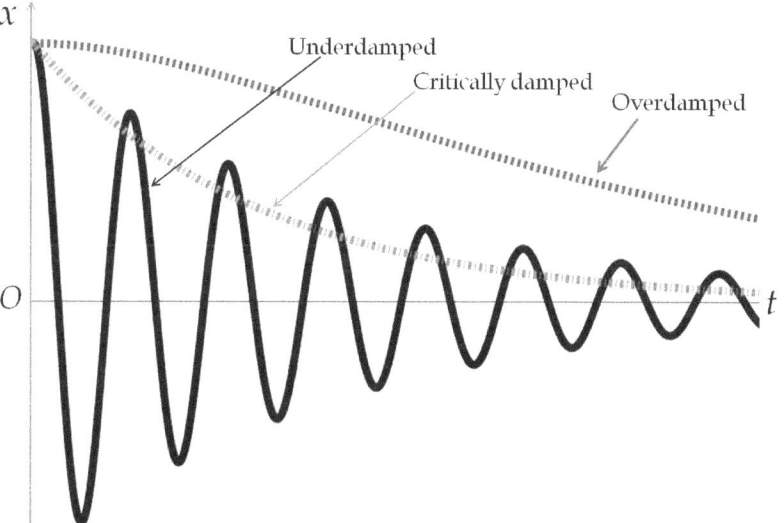

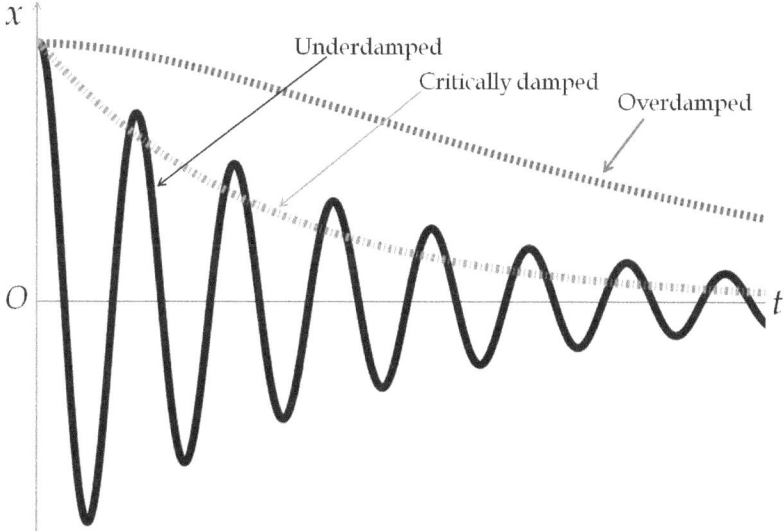

Fig. 3.16: Rear shock absorber and spring of a BMW R75/5 motorcycle. Photo credits: uploaded by Jeff Dean for Wikicommons.

Fig. 3.17: The displacement as a function of time for the under-damped, over-damped and critically damped cases. Only the under-damped case is an oscillator. To plot the figures, the following values were used: $x(0) = 1 \; cm$, $v(0) = 0$, $\beta = 1 \; rad/sec$, $\omega_1 = 10 \; rad/sec$, $\alpha = 0.2 \; rad/sec$.

The under-damped oscillator provides a physical meaning for the damping parameter β. We see from the graph in Fig. 3.18 that it takes a time of $1/\beta$ for the envelope of the oscillations to decrease by $1/e$ of its original value. The time to relax to $1/e$ of the original amplitude is called the **time constant** of the oscillator, which is also denoted by the Greek letter τ.

$$\tau = \frac{1}{\beta} \tag{3.80}$$

Therefore, the larger the damping constant, the shorter the time constant, and consequently, faster the damping.

3.4.2 Quality factor Q of an oscillator

Under-damped oscillators oscillate with decreasing amplitude and eventually come to rest. A better oscillator oscillates many more cy-

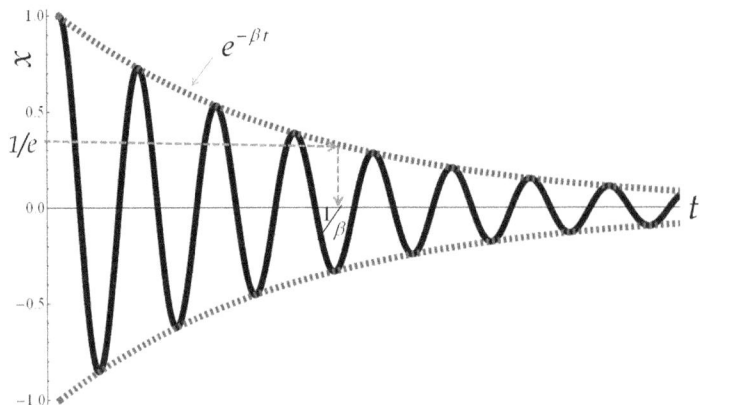

Fig. 3.18: The dynamics of under-damped oscillator shows oscillations with decreasing amplitude. The envelop of the decreasing amplitude is used to define the time constant for the under-damped oscillator. The figure shows that in time $t = 1/\beta$, the envelop decreased to by a factor of $1/e$, where e is the Euler's number with value $e = 2.71828....$

cles than a poorer oscillator before losing its energy. This important aspect of an oscillator is characterized by a quantity called the **Quality** or Q factor. The Q factor measures the persistence of oscillations of an oscillating system. A simple definition would be to divide the average energy in a particular cycle by the energy lost in the cycle, so that the less the energy is lost in a cycle, the more persistent the oscillations, and therefore higher the Quality of the oscillator. For technical reasons, we define the Q factor by dividing the energy at the beginning of a cycle to the energy lost in a fraction of the cycle.

$$Q = \frac{\text{Energy of the oscillator at the beginning of a cycle}}{\text{Energy dissipated per } (1/2\pi) \text{ of the next cycle}}. \quad (3.81)$$

The factor $1/2\pi$ of the cycle refers to per radian of the oscillation when we represent the motion of the oscillator by the motion of a fictitious particle in a circle so that covering 2π radians corresponds to one full cycle. One includes the factor $1/2\pi$ of the cycle in the definition to make the formulas later come out simpler. Although an exact calculation of the Q factor in terms of β and ω_0 of the oscillator is possible, it is not very illuminating. Instead, we will derive an approximate formula for Q of a lightly-damped oscillator defined by the following criterion.

$$\text{Lightly damped oscillator: } \beta << \omega_0. \quad (3.82)$$

First, note that the energy of a lightly-damped oscillator obeys the following equation (see the derivation below).

$$\frac{dE}{dt} = -2\beta E. \quad (3.83)$$

Therefore, the energy dissipated ΔE in a short time Δt is

$$\Delta E = \int_0^{\Delta t} \left| \frac{dE}{dt} \right| dt \approx \left| \frac{dE}{dt} \right| \Delta t = 2\beta E \Delta t. \qquad (3.84)$$

For Q, we need the energy loss in $1/2\pi$ of one cycle. Therefore, we choose Δt to be $1/2\pi$ times one time period T, which is

$$T = \frac{2\pi}{\omega_1} = \frac{2\pi}{\sqrt{\omega_0^2 - \beta^2}} \approx \frac{2\pi}{\omega_0}, \text{ since } \beta << \omega_0. \qquad (3.85)$$

Therefore, Q of a lightly damped oscillator is found to be

$$Q = \frac{E}{\Delta E \text{ for } 1/2\pi \text{ of one cycle}} \approx \frac{\omega_0}{2\beta}. \qquad (3.86)$$

Good oscillators such as tuning forks and guitar strings have Q values in the thousands. Laser cavities have much higher Q values, exceeding 10^7. Of course, the undamped oscillator has zero β, and hence infinite Q. There is no Q for the critically damped and over-damped cases since these systems do not oscillate.

Example 3.2. An under-damped harmonic oscillator. *A copper block of mass 1.5 kg is attached to a spring of stiffness 450 N/m and hung from a platform above a beaker that contains a thick liquid so that the block oscillates entirely in the liquid with a damping constant of 3.0 kg/s. (a) How many oscillations will the block make before its amplitude drops by 90%? (b) What is the Quality of this oscillator?*

Solution. a) We will first calculate the oscillation frequency ω_1 and the damping constant β for the damped oscillator. Since, the peak of successive cycles drops as $e^{-\beta t}$ we can find the time for the 90% drop in the amplitude using the value of β. Then, we will obtain the required number of oscillations by dividing the time required for the 90% drop by the time period of the oscillator.

$$\beta = \frac{b}{2m} = \frac{3.0 \text{ kg/s}}{2 \times 1.5 \text{ kg}} = 1.0 \text{ rad/sec.}$$

$$\omega_0 = \sqrt{\frac{k}{m}} = \sqrt{\frac{450 \text{ N/s}}{1.5 \text{ kg}}} = 17.32 \text{ rad/sec.}$$

Therefore, the angular frequency of oscillation is

$$\omega_1 = \sqrt{\omega_0^2 - \beta^2} \approx 17.3 \text{ rad/sec,}$$

which gives the time period for oscillations to be

$$T = \frac{2\pi}{\omega_1} \approx 0.363 \text{ sec.}$$

Now, we use the decay of the amplitude envelop to find the time for the amplitude to drop by 90%.

$$\frac{\text{Amplitude left}}{\text{Original Amplitude}} = 0.1.$$

Therefore,

$$e^{-\beta t} = e^{(-1)t} = 0.1 \implies t = 10 \text{ sec}.$$

Hence, the number of cycles in which the amplitude drops by 90% will be

$$\text{Number of cycles} = \frac{t}{T} = \frac{10 \text{ sec}}{0.362 \text{ sec}} = 27.5 \text{ cycles}.$$

(b) The Q factor of the oscillator is

$$Q \approx \frac{\omega_0}{2\beta} = \frac{17.32}{2(1)} = 8.66.$$

This is a poor oscillator, losing almost $2/3^{rd}$ of the amplitude in only 28 cycles.

3.4.3 Rate of change of energy for a lightly damped oscillator

A lightly-damped oscillator has much smaller damping constant (β) compared to the natural frequency (ω_0) of the oscillator.

$$\beta << \omega_0 \quad \text{(Lightly damped oscillator)}. \tag{3.87}$$

We will make use of this approximation to simplify formulas below. To find the rate at which energy is dissipated by a lightly damped oscillator, we can first calculate the energy from the sum of the kinetic and potential energies and then take the time derivative. Since a lightly damped oscillator is an under damped case we use the appropriate solution for the displacement $x(t)$.

$$x(t) = Ae^{-\beta t}\cos(\omega_1 t + \phi). \tag{3.88}$$

Note that we could have alternately used $x(t) = e^{-\beta t}\left[C_1 \cos(\omega_1 t) + C_2 \sin(\omega_1 t)\right]$, but, for the present calculation, the form given in Eq. 3.88 is better. The velocity is obtained by taking a time derivative of x:

$$v(t) = Ae^{-\beta t}\left[-\beta \cos(\omega_1 t + \phi) - \omega_1 \sin(\omega_1 t + \phi)\right]. \tag{3.89}$$

Putting these in the expression for energy we find the following for the energy at time t.

$$E = \frac{1}{2}mv^2 + \frac{1}{2}kx^2 \approx \frac{1}{2}m\omega_0^2 A^2 e^{-2\beta t}, \qquad (3.90)$$

where we have made use of the lightly damped approximation. Therefore, the energy of a lightly damped oscillator varies at the following rate.

$$\frac{dE}{dt} = -2\beta E. \qquad (3.91)$$

Note that the energy of the oscillator does not decrease at the same rate as the envelop of the amplitude: it takes a time equal to $1/2\beta$ for the energy to decrease by a factor $1/e$ while it takes twice as much time $1/\beta$ for the amplitude to decrease by a factor $1/e$. The reason for the energy to decrease twice as fast as the amplitude is that energy is proportional to the square of the amplitude. The same effect shows up in the rate at which sound fades compared to the rate at which a tuning fork loses vibrations. Since the intensity of sound generated by the vibrating tuning fork is related to the energy, the intensity of the sound will decrease faster than the amplitude of oscillations of the tuning fork itself.

3.5 DRIVEN OSCILLATOR

A damped oscillator will eventually come to rest if no additional energy is supplied to it. To supply energy to the oscillator you could apply any force that will do a net positive work on the oscillator. Among many forces that will do the job, a harmonically time varying force is of particular interest in physics for many applications. A harmonically varying force varies in time sinusoidally with a well-defined frequency ω_d.

$$\vec{F} = \vec{F}_0 \cos(\omega_d t), \qquad (3.92)$$

where F_0 is the amplitude of the force. Note that $\omega_d = 0$ corresponds to just a constant force, such as gravity; a constant force will do a positive work in one half of the cycle and the same amount of a negative work in the second half of the cycle, therefore a constant force will not change the energy of the oscillator over one complete cycle.

To be concrete, consider one-dimensional motion of a mass m attached to a spring subject to a spring force $\vec{F}_s$, a viscous force $\vec{F}_{\text{visc}}$, and an applied force $\vec{F}$ as shown in Fig. 3.19. The direction of the velocity at the instant is also shown in the figure so that you can see that the damping force and the velocity are in the opposite direction to each other. A simple physical realization of a damped driven oscillator is shown in Fig. 3.20.

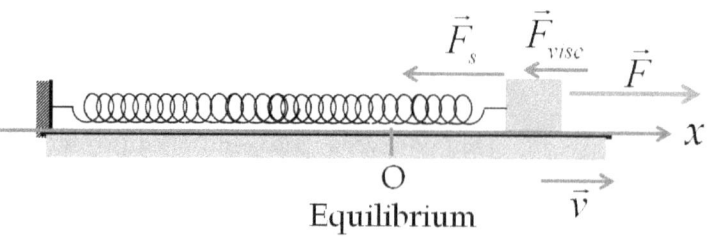

Fig. 3.19: Three forces act on a damped driven oscillator: $\vec{F}_s$ is the force by the spring, $\vec{F}_{\text{visc}}$ is the damping force, and $\vec{F}$ is the sinusoidal driving force. The velocity vector is shown to indicate the relative opposite directions of the damping force and the velocity vector.

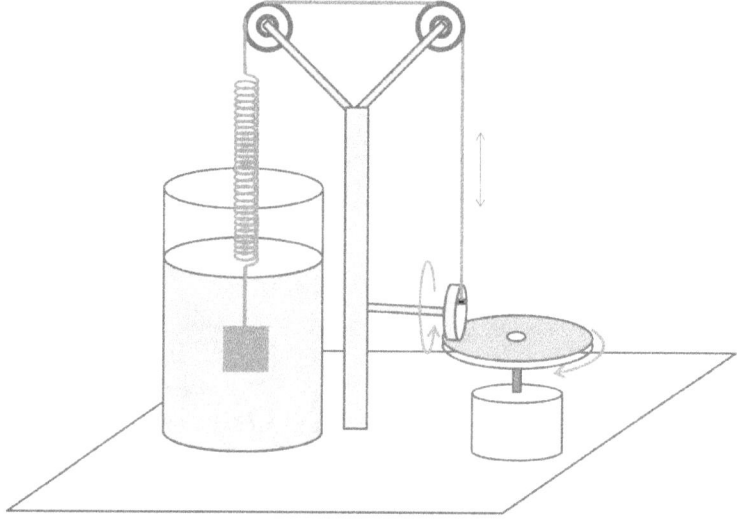

Fig. 3.20: A physical realization of the damped driven oscillator, called the "Texas Tower" has been developed by J. G. King at the Education Research Center, Massachusetts Institute of Technology. A variable motor drives the up and down motion of the block in the fluid sinusoidally. The fluid provides the damping force and the spring the restoring force.

The block in Fig. 3.19 moves in a straight line about an equilibrium point. We choose the origin of our coordinate system at the location of the equilibrium for the block, and the motion occurs on the x-axis as before. Then x-component of Newton's second law of motion gives the following equation of motion for the x-coordinate of the block.

$$m\frac{d^2x}{dt^2} = -kx - b\frac{dx}{dt} + F_0\cos(\omega_d t). \qquad (3.93)$$

Now we divide both sides of the equation by m and define the following composite parameters.

$$\omega_0 = \sqrt{k/m}, \text{ the natural frequency,} \qquad (3.94)$$

$$\beta = b/2m, \text{ the damping constant,} \qquad (3.95)$$

$$D = F_0/m. \qquad (3.96)$$

The quantity D is the acceleration of the block if only the driving force were to act on the system at zero driving frequency. Using the new parameters the equation of motion can be rewritten as follows,

$$\frac{d^2x}{dt^2} + 2\beta\frac{dx}{dt} + \omega_0^2 x = D\cos(\omega_d t), \tag{3.97}$$

whose solution describes the motion of the block. The differential equation obtained here is considerably more difficult to solve than the ones we had encountered for the undamped and the viscously damped oscillators.

Since the block is being forced to oscillate at the driving frequency, ω_d, we expect that in the long run, the block should oscillate at this frequency. The solution for a harmonically driven damped oscillator does exhibit this behavior when $t >> 1/\beta$. While in the short time of turning on the driving force, the x-coordinate of the block is a complicated function of time, the long-term behavior is a simple harmonic motion at the driving frequency. The complete solution of Eq. 3.97 that describes the short-term solution accurately is called the **transient solution**. The long-term solution, where the transients have died out and the block executes a simple harmonic motion, is called the **steady state solution**.

When the block reaches a steady state, it just oscillates at the frequency of the driving force with an amplitude and phase that depend upon the frequency of the driving force, among other parameters. We will state here the steady state solution without actually solving the equation of motion. Even though we do not solve the equation, various aspects of the solution are important for us to discuss here. The steady state solution of Eq. 3.97 can be written either as a cosine or a cosine with a phase constant or a mixture of sine and cosine functions. To be specific, let us write the solution as a cosine with a phase constant.

$$x_s(t) = A\cos(\omega_d t - \delta). \tag{3.98}$$

Other ways of writing the same answer are: $A'\sin(\omega_d t - \delta')$ and $C_1\cos(\omega_d t) + C_2\sin(\omega_d t)$. Recall that the driving force is $F_x = F_0\cos(\omega_d t)$. Therefore, the phase constant δ of the displacement x_s represents the phase lag of the displacement with respect to that of the driving force as shown by using the circular motion analogy for the phase constant in Fig. 3.21.

The value of the phase constant says the relative position of the displacement and the driving force in their own cycles. For instance, if $\delta = 0$, then the displacement of the block and the driving force are synchronized in the sense that the when displacement is at the peak then so is the driving force and when the displacement is at the

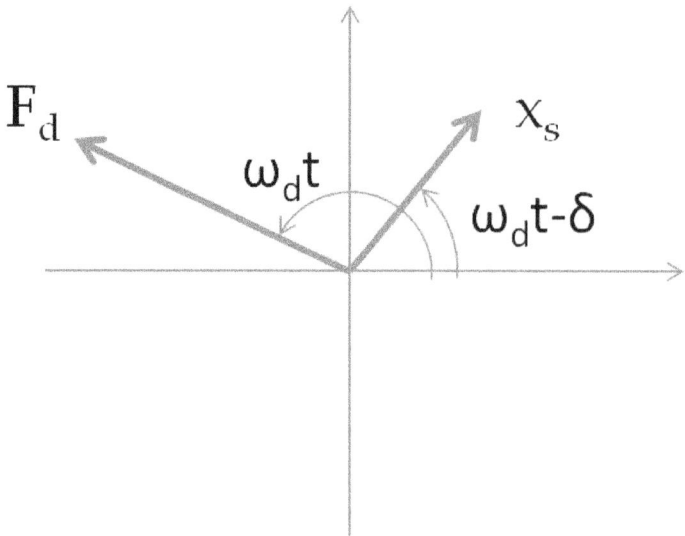

Fig. 3.21: Phase lag of displacement with respect to the driving force shown here using the circular motion analogy for phase given earlier in the chapter.

trough so is the driving force. If $\delta = \pi$, the displacement of the block is pointed in opposite direction to that of the driving force, and if $\delta = \pi/2$, then the displacement is quarter cycle behind the force.

You can solve for the two unknowns A and δ in the solution given in Eq. 3.98 by putting the solution into Eq. 3.97. The algebra is tedious but do-able. You should attempt to show that one would get the following expressions for the amplitude and phase constant.

$$A = \frac{D}{\sqrt{(\omega_0^2 - \omega_d^2)^2 + (2\beta\omega_d)^2}} \tag{3.99}$$

$$\tan\delta = \frac{2\beta\omega_d}{\omega_0^2 - \omega_d^2} \tag{3.100}$$

3.5.1 Steady state and resonance

In the steady state, the oscillator oscillates between $x = -A$ and $x = A$ at the angular frequency ω_d. The solution for A given above shows that the amplitude A of the motion varies with the strength of the force through $D = F_0/m$, as you would expect, but also, more importantly, with the frequency ω_d of the harmonic driving force. That is, you can get different amplitudes of oscillation for the same magnitude of the force F_0 if you apply the fore at a different frequency.

Fig. 3.22 shows the variation of A with respect to the driving frequency. The figure illustrates visually that when you vary the frequency ω_d of the driving force, the steady state amplitude of the

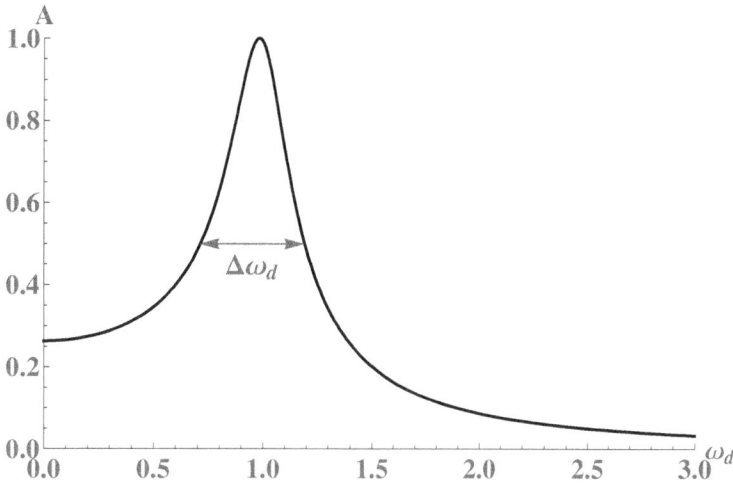

Fig. 3.22: The amplitude of the steady state oscillations as a function of driving frequency. Here the amplitude A divided by the amplitude at the resonance A_R is plotted along y and driving frequency in units of ω_0 is plotted along x-axis for an oscillator of damping constant $\beta = 0.1\omega_0$. The width at half-height $\Delta\omega_d$ characterizes the sharpness of the peak.

oscillations of the block would change. We see that the amplitude takes its maximum value at a particular value of ω_d. This special frequency is called the **resonance frequency**, which we will denote by ω_R. To find the expression of the resonance frequency in terms of other quantities, we can apply Calculus to the function $A(\omega_d)$ and set its first derivative to zero and solve for ω_d. This ω_d will equal the resonance frequency ω_R.

$$\left.\frac{dA}{d\omega_d}\right|_{\omega_d=\omega_R} = 0. \tag{3.101}$$

Rather than do the calculation indicated in this equation, we note that the extrema of A with respect to ω_d will occur at the same place as the extrema of the quantity inside the radical in the denominator of A given in Eq. 3.99. This provides a much simpler calculation for the same result.

$$\left.\frac{d[(\omega_0^2 - \omega_d^2)^2 + (2\beta\omega_d)^2]}{d\omega_d}\right|_{\omega_d=\omega_R} = 0. \tag{3.102}$$

We can solve this equation for ω_R. Keeping only the positive root, we find the following expression for the resonance frequency.

$$\boxed{\omega_R = \sqrt{\omega_0^2 - 2\beta^2}.} \tag{3.103}$$

The frequency $\omega_d = \omega_R$ is not only an extremum of $A(\omega_d)$ but also its maximum as you can verify by finding the sign of the second derivative of A with respect to ω_d for $\omega_d = \omega_R$. Therefore, the oscillator will vibrate with the largest amplitude if it driven at this frequency.

Equation 3.103 shows that, normally, the resonance frequency is below the natural frequency of the oscillator. However, for a lightly-damped oscillator ($\omega_0 \gg \beta$) the resonance frequency will be very near its natural frequency, $\omega_0 \equiv \sqrt{k/m}$.

$$\text{Lightly damped oscillator: } \omega_R \approx \omega_0. \qquad (3.104)$$

Therefore, in a lightly-damped oscillator, an experimental determination of the resonance frequency gives a good indicator of the natural frequency of the system. What happens if the oscillator is driven at the resonance frequency? We can find the answer by directly putting $\omega_d = \omega_R$ in Eqs. 3.99 and 3.100. Let us denote the values of the amplitude and phase lag at the resonance by A_R and δ_R respectively.

$$A_R \equiv A(\omega_d = \omega_R) = \frac{D/2\beta}{\sqrt{\omega_0^2 - \beta^2}} \qquad (3.105)$$

$$\tan \delta_R \equiv \delta(\omega_d = \omega_R) = \frac{2\beta\omega_R}{\omega_0^2 - \omega_R^2} = \frac{\pi}{2}. \qquad (3.106)$$

Therefore, at the resonance, the position of mass and the driving force are 90-degrees out of phase with each other. This says that at the resonance, while the driving force varies as $\cos(\omega_R t)$, the position of the block varies as $\sin(\omega_R t)$.

$$x_s\big|_{\omega_d = \omega_R} = A_R \, \cos\left(\omega_R t - \frac{\pi}{2}\right) = A_R \, \sin\left(\omega_R t\right). \qquad (3.107)$$

The occurrence of the largest amplitude for the oscillator at the resonance implies that the external agent is able to transfer energy to the oscillator most effectively under these conditions. This is seen quite readily in a swing when you try to make the person in the swing to oscillate at larger amplitudes. You must time the push on the swing properly in order to make it oscillate with a large amplitude.

The resonance phenomenon can also cause serious problems in physical structures. Gusty winds can drive buildings and other structures to swing, and if the gusts have a frequency component that matches with the resonance frequency of the structure, the resonance can then drive the structure to larger amplitude oscillations. A famous disaster as a result of gusty winds happened in 1940 when Tacoma Narrows Bridge in Washington state broke under large amplitude oscillations, which may have been caused by resonance. To prevent the resonance to take place, engineers use dampers to critically damp large buildings.

The resonance phenomenon is also important in the microscopic world. For instance, we use the resonance of nuclear moments of protons in the nuclear magnetic resonance (NMR) to investigate the structure of matter and for medical diagnostic applications. Similarly, in lasers we use resonance to amplify the electromagnetic field inside the cavity.

3.5.2 Role of Damping in Resonance

We have introduced the Quality factor $Q = \omega_0/2\beta$ to characterize the effect of damping on harmonic oscillations. Therefore, let us rewrite our formulas for the amplitude A and the phase lag δ in terms of the quality factor Q by substituting β in terms of Q using $\beta = \omega_0/2Q$ in the Eqs. 3.99 and 3.100.

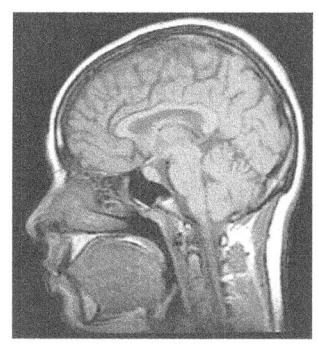

Fig. 3.23: Nuclear magnetic resonance is used in the Magnetic Resonance Imaging to look inside human body. The inside picture of a brain tells activities in various parts of the brain. Picture credits: Wikicommons.

For calculations, it is convenient to define a dimensionless frequency by dividing the frequency ω_d of th driving force by the natural frequency ω_0 of the oscillator. We introduce a dimensionless parameter $\Omega = \omega_d/\omega_0$ that will vary with ω_d.

$$\Omega \equiv \frac{\omega_d}{\omega_0}$$

It is helpful in the calculations to replace ω_d by $\Omega\omega_0$. We will see that ω_0 will cancel out from the expressions. After some algebra, which is left as an exercise for the student and is highly recommended to the student, we find that

$$A = \frac{D}{\omega_0^2} \frac{1}{\sqrt{(1-\Omega^2)^2 + \Omega^2/Q^2}}, \tag{3.108}$$

$$\tan\delta = \frac{\Omega/Q}{1 - \Omega^2}, \tag{3.109}$$

The resonance frequency in the units of ω_0 takes the following form.

$$\Omega_R \equiv \frac{\omega_R}{\omega_0} = \sqrt{1 - \frac{1}{2Q^2}}. \tag{3.110}$$

We plot Eqs. 3.108 and 3.110 in Fig. 3.24, and Eqn. 3.109 in Fig. 3.25. The resonance peaks of the amplitude versus frequency shows that the resonance becomes taller and sharper with the lowering of damping which is the same as increasing of the Quality factor Q. The resonance peak does not coincide with the natural frequency ω_0 as seen in Fig. 3.24(b) but approaches ω_0 as seem from $\Omega_R \to 1$ when $Q \to \infty$. For good oscillators with low damping we can usually take the resonance frequency to be equal to the natural frequency.

Figure 3.25 shows the variation of phase lag for different Quality factor. The phase lag goes from zero to π when the driving frequency

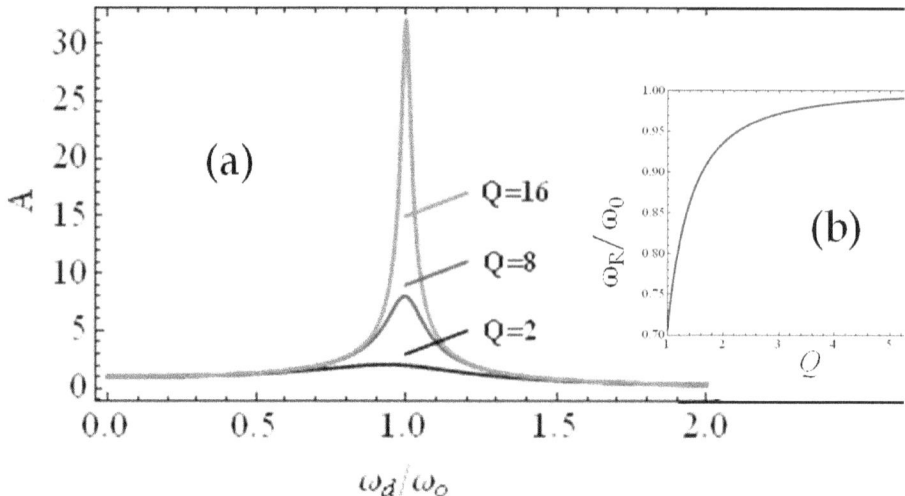

Fig. 3.24: The Resonance for different Q. (a) Plot of amplitude A as a function of driving frequency for $Q = 2$, 4, and 32. (b) The resonance frequency ω_R, here the ratio $\Omega_R = \omega_R/\omega_0$ as a function of Q shows that resonance frequency tends to the natural frequency ω_0 as Quality Q of the oscillator rises.

goes from low frequencies to high frequencies with a transition at the resonance frequency. The transition at resonance frequency becomes sharper for higher Q oscillators. For an undamped oscillator, the transition is a step function.

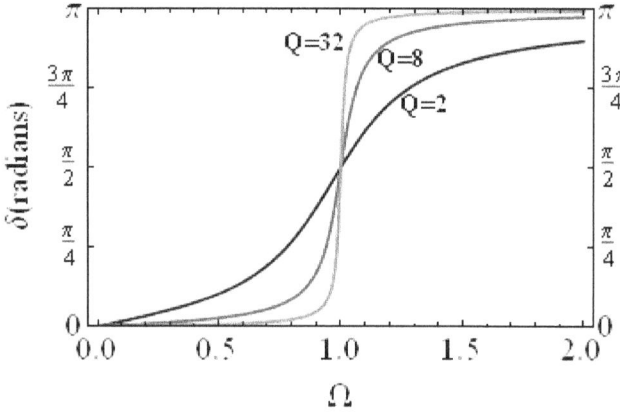

Fig. 3.25: The phase lag as a function of driving frequency for different Q.

3.5.3 Power of the Driving Force

The driving force must do work to overcome the dissipation of energy due to the friction of the damping force in order to drive a steady motion of the oscillator. The power expended by the driving force is equal to the driving force times the velocity of the oscillator. We use the situation in the steady state and the steady state velocity to obtain the following instantaneous power $P(t)$ by the driving force.

$$
\begin{aligned}
P(t) &= \vec{F}_d \cdot \vec{v} \\
&= F_0 \cos(\omega_d t) \frac{dx_p}{dt} \\
&= F_0 \cos(\omega_d t) \frac{d}{dt} \left[A \cos(\omega_d t - \delta) \right] \\
&= -F_0 \omega_d A \cos(\omega_d t) \sin(\omega_d t - \delta) \\
&= \frac{F_0 \omega_d A}{2} \left[\sin \delta + \sin \delta \cos(2 \cos \omega_d t) \right. \\
&\qquad \left. - \cos \delta \sin(2 \cos \omega_d t) \right] \quad (3.111)
\end{aligned}
$$

What is important is power expended by the force over a full cycle - this is the energy that the driving force must supply per cycle to maintain the steady motion of the oscillator. The average power P_{av} is obtained by integrating Eq. 3.111 over one cycle. The integration over the cosine and sine in Eq. 3.111 will give zero leaving only the contribution from the constant term.

$$
\begin{aligned}
P_{av} &= \frac{\omega_d}{2\pi} \int_0^{2\pi/\omega_d} P(t) dt \\
&= \frac{1}{2} F_0 \omega_d A \sin \delta, \quad (3.112)
\end{aligned}
$$

The average power is also a function of the driving frequency. Let us write the full expression for the average power using explicit expressions for $A(\omega_d)$ and $\delta(\omega_d)$. After much algebra, it can be shown that

$$
P_{av} = \left(\frac{\beta F_0^2}{m} \right) \frac{\omega_d^2}{\left(\omega_0^2 - \omega_d^2 \right)^2 + (2\beta\omega_d)^2}. \quad (3.113)
$$

Unlike the amplitude A of the displacement of the oscillator, the average power of the driving force is maximum when $\omega_d = \omega_0$ regardless of the value of the damping constant as shown in Fig. 3.26. Often the resonance of power is taken to mean as the resonance phenomenon rather than the resonance of the amplitude A. Note that there are differences in the resonance of the average power of the driving force and the resonance of the amplitude of the displacement of the oscillator from the equilibrium. The resonance of the power is important experimentally, since in many experimental settings, it is easier to measure power than the displacement.

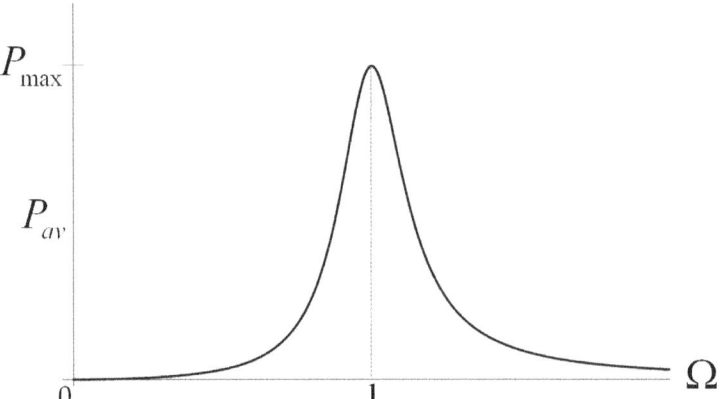

Fig. 3.26: The power curve of driven oscillator. The resonance of power occurs at $\Omega = \omega_d/\omega_0 = 1$. The maximum power at the resonance has the value $P_{\max} = F_0^2/4\beta m$. The width of the curve at half height of the peak, denoted as $\Delta\Omega_{1/2}$ is approximately 2β. and the width for a lightly damped oscillator is determined by the damping constant $2\beta/\omega_0$.

3.6 EXERCISES

Simple Harmonic Oscillator

Ex 3.1. A harmonic oscillator of mass 200 grams is attached to a spring of spring constant 100 N/m. Find the angular frequency, frequency and time period.

Ex 3.2. Consider a simple harmonic oscillator of mass m, amplitude A and frequency f. (a) What fraction of the energy of the oscillator is in the kinetic energy when the oscillator's displacement is half the amplitude? (b) What fraction of the energy of the oscillator is in the potential energy when the oscillator's displacement is half the amplitude? (c) Where in the cycle of an oscillation does the oscillator have the lowest speed? Why? (d) Where in the cycle of an oscillation does the oscillator have the largest speed? Why?

Ex 3.3. Two oscillators of masses 200 g and 400 g oscillate with the same frequency. They move such that their positions from their corresponding equilibrium points are given by the following two functions (t in seconds, x in cm) x_1 and x_2 respectively.

$$x_1 = 2\cos(t)$$
$$x_2 = 2\sin(t)$$

(a) What are the angular frequencies, amplitudes and phase constants of the two oscillators? (b) What were their positions and velocities at the initial time $t = 0$? (c) Plot the positions of the two oscillators versus time on the same graph, and interpret which oscillator is ahead, and by how much. (d) What are the kinetic and potential energies of the two oscillators at t=0?

Ex 3.4. Two oscillators of masses 100 g and 400 g oscillate with the same frequency. They move such that their positions from their corresponding equilibrium points are given by the following two functions (t in seconds, x in cm).

$$x_1 = 2\cos(t)$$
$$x_2 = 2\cos(t - \pi)$$

(a) What are the angular frequencies, amplitudes and phase constants of the two oscillators? (b) What were their positions and velocities at the initial time $t = 0$? (c) Plot the positions of the two oscillators with time and interpret which oscillator is ahead, and by how much. (d) What are the kinetic and potential energies of the two oscillators at $t = 0$?

Ex 3.5. A block of mass 5 kg is hung from a spring of the original length is 50 cm and the spring constant 100 N/cm. As a result of the weight of the block, the spring stretches to a different length at equilibrium. The spring is attached to a ceiling at a height 2 m from the ground. The block is then hit with a hammer giving it an instantaneous velocity of 4 cm/s downward. (a) When the block is hanging in equilibrium, where is the equilibrium position of the block with respect to the floor? (b) Pick the zero reference points for the potential energies due to the spring force and that due to the gravity, and determine the potential energy of the block at the time it was hit by the hammer? (c) What is the kinetic energy of the block immediately after being hit by the hammer? (d) How much work did the hammer do? (e) What is the frequency of the oscillations of the block? (f) What is the phase constant of the motion of the block? (g) Suppose the y-axis is pointed up with the origin at the equilibrium when the block is hanging with zero net force on the block, write a function $y(t)$ for the y-coordinate of the block. (h) Find the turning points in the motion of the block.

Ex 3.6. An oscillator of frequency 20 cycles per second starts 3 cm from the equilibrium with an initial velocity of 10 cm/s pointed towards the equilibrium point. Draw a figure, showing the origin and the x-axis along the motion of the oscillator and find $x(t)$ for the oscillator.

Ex 3.7. An oscillator of frequency 20 cycles per second starts 3 cm from the equilibrium with an initial velocity of 10 cm/s pointed away from the equilibrium point. Find the displacement as a function of time.

Ex 3.8. The solution of a Simple Harmonic Motion can be written in three ways:

$$x(t) = C_1 \, \cos(\omega t) + C_2 \, \sin(\omega t)$$
$$= A \cos(\omega t + \phi_c)$$
$$= B \sin(\omega t + \phi_s)$$

Find the relations among C_1, C_2, A, ϕ_c, and ϕ_s.

Ex 3.9. Find C_1 and C_2 in terms of x_0, v_{0x} and ω when $x(t) = C_1 \, \cos(\omega t) + C_2 \, \sin(\omega t)$, where $x_0 = x(0)$ and $v_{0x} = v_x(0)$.

The Damped Harmonic Oscillator

Ex 3.10. Decide what type of damping is in the following oscillators: (a) $m = 0.2$ kg, $k = 10$ N/m, $b = 6$ kg/s (b) $m = 0.2$ kg, $k = 20$ N/m, $b = 8$ kg/s (c) $m = 0.8$ kg, $k = 32$ N/m, $b = 8$ kg/s (d) $m = 0.8$ kg, $k = 80$ N/m, $b = 16$ kg/s

Ex 3.11. If $x(0) = 1$, and $v_x(0) = 0$ for the oscillators in Exercise 3.10, find $x(t)$ for each oscillator.

Ex 3.12. Find the Q factor for the following underdamped oscillators. Here x is in cm and t in sec.

(a) $x(t) = 2 \exp(-0.1t) \cos(2\pi t)$
(b) $x(t) = 2 \exp(-0.02t) \cos(t)$
(c) $x(t) = 2 \exp(-0.25t) \cos(200\pi t)$
(d) $x(t) = 2 \exp(-0.015t) \cos\left(2000\pi t + \frac{\pi}{2}\right)$

Ex 3.13. The position of a 250-gram lightly damped oscillator is given by the following function of time, $x(t) = 2 \exp(-0.01t) \cos(2\pi t)$, where t is in seconds and x in meters.

(a) Plot x vs t.
(b) How long does it take for the envelop of oscillations to drop by $\frac{1}{e}$?
(c) How long does it take the envelop to drop by a factor $\frac{1}{e^2}$?
(d) What is the Q factor of the oscillator?
(e) What is the rate at which the energy of the oscillator is dissipated

at $t = 0$?

(f) What is the frequency of oscillations?

(g) How many oscillations will the oscillator make before 90% of the energy is dissipated?

Ex 3.14. A block of mass m attached to a spring of stiffness constant k is hung vertically in a fluid which damps the motion of the mass. The damping force is proportional to the speed with a constant of proportionality b. Ignore the force of buoyancy. The block is pulled a distance A from the equilibrium position so that the spring stretches by A, and released from rest. Find the subsequent motion by deriving the expressions for the displacement and velocity at an arbitrary time.

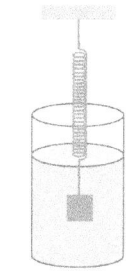

Fig. 3.27: Exercise 3.14

The Forced Harmonic Oscillator

Ex 3.15. An underdamped oscillator with $\beta = \frac{\omega_0}{10} = 2$ rad/sec is driven by a harmonic driving force $\vec{F}_d$. The oscillator oscillates along the x-axis. The driving force has the following x-component: $F_x = (5N) \cos(\omega_d t)$. (a) Find the steady state amplitude of the oscillator when the driving frequency is equal to the resonance frequency. (b) By what factor the amplitude drops compared to the amplitude at the resonance, if the driving frequency is equal to $0.1\omega_R$. (c) At what frequencies would the amplitude be half of the amplitude at the resonance?

Ex 3.16. An underdamped oscillator with $\beta = \frac{\omega_0}{10} = 2$ rad/sec is driven by a harmonic driving force $\vec{F}_d$. The oscillator oscillates in x-axis. The driving force has the following x-component: $F_x = (5N) \cos(\omega_d t)$. (a) Find the phase difference between the driving force and the displacement, if the driving frequency is equal to (i) $0.1\omega_R$, (ii) $0.9\omega_R$, or (iii) $1.1\omega_R$. (c) At what frequency would the phase difference be $-\frac{\pi}{4}$ radian?

Ex 3.17. A block of mass 300 grams is attached to a spring of spring constant 100 N/m. The block moves in an environment that damps its motion. The damping force is proportional to the speed of motion of the mass with the constant of proportionality given as 0.1 N.s/m. A sinusoidal driving force acts on the block with amplitude 10 N and a variable frequency. Find the following.

(a) The natural frequency of the oscillator

(b) The type of damping of the oscillator - under, over or critical

(c) The Q factor of the oscillator

(d) The resonance frequency of the oscillator

(e) The amplitude and phase of the oscillator in the steady state when the frequency of the driving force are: i. $0.25\ \omega_R$, ii. $0.5\ \omega_R$, iii. $0.75\ \omega_R$, iv. ω_R, v. $1.25\ \omega_R$, vi. $1.5\ \omega_R$, vii. $1.75\ \omega_R$, viii. $10\ \omega_R$.

Ex 3.18. Consider a one-dimensional damped oscillator of mass $m = 0.5$ kg, $\omega_0 = 10$ rad/sec, and $\beta = 1$ rad/sec. Suppose the oscillator can oscillate along the x-axis. A sinusoidal force with the x component $F_x = 100$ N $cos(\omega t)$ acts on the oscillator. (a) What should be the condition on time t so that the steady state for the oscillator can be assumed? (b) Suppose the steady state has been reached and the displacement of the oscillator can be written as $x = A_c \cos(\omega t) + A_s \sin(\omega t)$. Find the amplitudes A_c and A_s as a function of driving frequency ω. (c) Find the amplitude of the steady sate oscillations. (d) Find the phase lag? (e) What is the expression for the instantaneous power of the driving force? (f) Find the average power as a function of driving frequency. (g) At what frequency is the power maximum? (h) At what frequency is the amplitude of the oscillator's motion maximum? (i) What is the width at half height of the power versus driving frequency plot?

Ex 3.19. A harmonically driven lightly damped oscillator of mass 200-grams has a resonance frequency of 100 Hz. At the resonance frequency its amplitude is 750 N/kg. The applied force has a peak amplitude of 2 N. (a) Find the quality factor (Q) of the oscillator. (b) What is the approximate value of the damping constant b of the damping force?

3.7 PROBLEMS

P 3.1. The mass of a body on Earth is easily found by using a spring balance which measures weight mg of the body. In outer space this method is not available for determining mass of a body. (a) Design a method for finding mass based on simple harmonic oscillator. (b) Describe how you will find the mass of an astronaut in space.

P 3.2. A plank of mass 10-kg rests on four identical springs of spring constant k. (a) What can be the largest spring constant if the plank is not to vibrate at greater than 1 Hz frequency? (b) If 20 kg of lead is put on the plank, what will be the new frequency of vibration? Ans: (a) 99 N/m; (b) 0.5 Hz.

P 3.3. A disk of mass m and radius R is suspended from a thin string of torsion constant k. The string is fixed to the center of disk and oriented perpendicular to the disk. When the disk is rotated by a small angle about the equilibrium, there is a restoring torque due to twist in the string, which tends to bring the disk back to the equilibrium. Find the frequency of small oscillations about the equilibrium. Ans: $\frac{1}{2\pi R}\sqrt{2kM}$.

P 3.4. A bullet of mass m and speed v_0 is fired horizontally on a block of mass M attached to a spring of spring constant k whose other end is fixed to a wall. The direction of the velocity of the bullet is along the length of the spring. Upon impact, the bullet is embedded in the block. (a) Ignoring damping, find the amplitude of the resulting harmonic motion. (b) Compare the combined mechanical energy of the bullet and the block together after the impact to that of their mechanical energy before the impact. Ans: (a) $\frac{mv_0}{\sqrt{k(M+m)}}$; (b) $\frac{E_{\text{after}}}{E_{\text{before}}} = \frac{m}{M+m} < 1$.

P 3.5. A mass m rests over a block of mass M which rests on a frictionless table. A spring of spring constant k is attached to M. The two-mass assembly is pulled so that the spring stretches by a distance A and then released from rest. If the stretching is greater than a critical value A_0, the block of mass m slides on the block M. What is the coefficient of static friction between m and M?

P 3.6. Two springs of spring constants k_1 and k_2 are attached on the two opposite sides of a block of mass m and the free ends of the springs are attached to two fixed supports such that the springs are taut and stretched. The block rests on a frictionless flat horizontal surface and can move along the line of the two springs. When the block is pulled a little from the equilibrium position in the line of the springs and released from rest, the block executes a simple harmonic motion whose frequency depends on the spring constants of the two springs and the mass of the block. (a) By looking at forces on the block at an arbitrary point in time, find the equation of motion of the block. (b) From the equation of motion, show that the angular frequency of oscillation is given by $\omega = \sqrt{\omega_1^2 + \omega_2^2}$, where $\omega_1^2 = k_1/m$ and $\omega_2^2 = k_2/m$. (c) Deduce the formula for the frequency by examining the expression for the energy of the system.

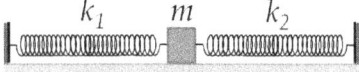

Fig. 3.28: Problem 3.18

P 3.7. Show that the system of two springs of spring constants k_1 and k_2 connected in parallel to an object has an effective spring constant equal to $k_{\text{eff}} = k_1 + k_2$.

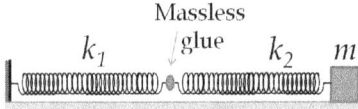

Fig. 3.29: Problem 3.8

P 3.8. Two springs of spring constants k_1 and k_2 are glued with a light but strong glue. One end of the combination is attached to a fixed wall and a block of mass m is attached to the other end and the block is placed on a frictionless table as shown in Fig. 3.29. (a) Write the equation of motion of the block at an arbitrary time. (b) Write the equation of motion of the glue at the junction of the two springs and simplify the equation by setting the mass of the glue to zero. Assume mass of the glue negligible so that $m\vec{a}$ of the glue can be taken to be zero. (c) Combine the two equations to show that the frequency of oscillation of the block is given by $\omega = \frac{\omega_1 \omega_2}{\sqrt{\omega_1^2 + \omega_2^2}}$, where $\omega_1^2 = k_1/m$ and $\omega_2^2 = k_2/m$. (d) Deduce the formula for the frequency by examining the expression for the energy of the system.

P 3.9. Show that the system of two springs of spring constants k_1 and k_2 connected in series to an object has effective spring constant that obeys $\frac{1}{k_{\text{eff}}} = \frac{1}{k_1} + \frac{1}{k_2}$.

P 3.10. A U-tube of cross-section area A and distance w between the necks, is filled by a liquid of density ρ up to a height h_0. At equilibrium the level of the liquid on the two sides is at the same height h_0 from the ground. The liquid on one side is pushed in by a small height Δh and let go. It is then observed that the liquid executes a simple harmonic motion about the equilibrium level h_0. For this problem ignore the effects of resistance and viscosity. Use a coordinate system whose y-axis is pointed up and the origin is at the bottom point of the U-tube to answer the following questions. (a) Find an expression for the potential energy of water at an arbitrary time. Express your answer in terms of the y-coordinate of the level in one of the arms. (b) Find an expression for the kinetic energy of the liquid at an arbitrary time. Note that all liquid in the U-tube will be moving. Express your answer in terms of dy/dt, the rate at which the level will be changing. (c) Find the frequency of oscillation.

P 3.11. The power input to an oscillator by the applied force depends on the velocity $\vec{v}$ of the oscillator and the applied force $\vec{F}$ at that instant as given by the defining equation for the instantaneous power, $P(t) = \vec{F} \cdot \vec{v}$. By integrating over a period and dividing by the period, we obtain an average power of the force. (a) Find a formula for the average power delivered to a sinusoidally driven oscillator in the steady state. (b) From the formula in (a), deduce the formula for the resonance frequency of average power.

P 3.12. The front suspension of a car has a natural frequency of 0.5 Hz. At the time the car was built, according to specifications, the shock absorbers of the car provide critical damping. With time, the shock absorbers get worn out so that they no longer provide the critical damping. The car then acts as an under-damped oscillator: when the car goes over a bump, it oscillates through many cycles. From measurements on the damped oscillations of the car, you calculate that $\beta = 0.4$ rad/sec. When the car is driven on a bumpy road with bumps placed at regular intervals of 50 m, the car shakes violently when driven at a particular speed. Find this critical speed that the driver must avoid.

P 3.13. To determine the value of acceleration due to gravity from a physical pendulum one needs to determine the time period and the moment of inertia of the pendulum. However, the moment of inertia is usually difficult to determine experimentally. The Kater's pendulum is a special physical pendulum that removes the necessity of knowing the moment of inertia by using the periods from two suspension points.

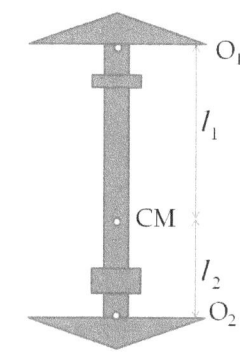

Fig. 3.30: A Kater's pendulum.

The **Kater's pendulum** consists of a long metallic rod with a slideable weight on the rod. The pendulum has two knife edges at two ends that are used to suspend the pendulum from a support with a hole over which the knife edge rests. By adjusting the pivot points and the mass distribution over the rod, it is possible to obtain a configuration such that the time periods of oscillations about the two suspension points O_1 and O_2 are equal. Let k be radius of gyration of the pendulum, l_1 and l_2 the distances from the center of gravity to the suspensions O_1 and O_1 respectively. Let T_1 and T_2 be the time periods of oscillations when Kater's pendulum is hung from knife edges O_1 and O_2 respectively.

(a) Find the time periods T_1 and T_2 in terms of l_1, l_2, k, and g. (b) Prove that when the time periods are equal, the period is given by $T = 2\pi\sqrt{\frac{L}{g}}$, where $L = l_1 + l_2$, the distance between the two knife edges.

P 3.14. The potential energy of a particle of mass m moving along x-axis is given as $U(x) = x(x-1)^2$. (a) Plot the potential energy as a function of x. Discuss the types of motion that a particle will execute for various values of energy of the particle. (b) Find the location of the minimum of the potential energy function. (c) Find the angular frequency of small oscillation about the minimum of the potential. (d) Are there any restrictions on the displacement for the particle to oscillate about the minimum? Explain.

P 3.15. The potential energy of a particle of mass m moving along x-axis is given as $U(x) = x^2(x-1)^2$. (a) Plot the potential energy as a function of x. Discuss the types of motion that a particle will execute for various values of energy of the particle. (b) Find the location of the minima of the potential energy function. (c) Find the angular frequency of small oscillation about the minimum of the potential. (d) Are there any restrictions on the displacement for the particle to oscillate about the minimum? Explain.

Chapter 4

COUPLED VIBRATIONS AND WAVE MOTION

Contents

To most people, waves conjure up images of water waves striking a lake shore or an oceanfront. When you throw a stone in water you can clearly see the circular water waves traveling outward from the point of impact (Fig. 4.1). There are many types of waves in nature, e.g. sound waves, seismic waves, waves on a string, electromagnetic waves, etc., just to a name a few of the commonly encountered ones. Despite a tremendous variety, all waves share some common characteristics. Waves involves vibration of either the particles of the medium or some property, and transport energy and momentum over space without an actual transport of material.

For instance, when you throw a stone in a pond, the energy in the stone goes in starting vibrations of water molecules at the point of impact of the stone. The vibrations then travel outward as a wave which transport the energy to the shore. As the wave pattern travels through water, water molecules oscillate up and down about an equilibrium, as seen in the motion of a light object such as a leaf at the surface.

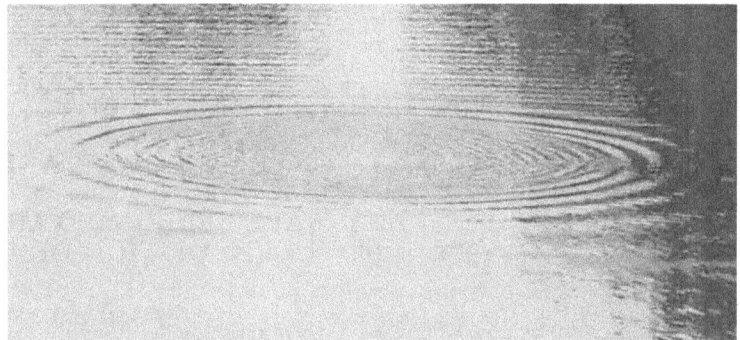

Fig. 4.1: Waves in lake generated by dropping a stone.

There is a close connection between waves and oscillations. When a sound wave travels through air, the air particles oscillate to and fro about their equilibrium point in the direction of the wave. The oscillating particles of air are coupled to each other so that oscillation of one layer of particles can lead to the oscillation of the neighboring particles, thereby transferring energy to their neighbors. This process repeats in the medium. The mechanism does not require the transport of air particles themselves, but just the transfer of energy and momentum of the particles through local oscillations. The pattern of oscillations make up a wave. In this chapter we will study common characteristics of all waves using mainly the mechanical waves on a taut string as an example.

4.1 COUPLED VIBRATIONS OF TWO MASSES

In the last chapter we have studied oscillations of one mass about an equilibrium. Wave phenomena arises as a result of the coupling of vibratory motions of different parts of a system. To develop a feel for the consequences of coupling of vibrations, we will study a simple system consisting of two blocks of masses m_1 and m_2 that are connected by a spring and supported on the two sides by two other springs as shown in Fig. 4.2. We will study the one-dimensional motion of the coupled masses analytically by choosing the x-axis to coincide with the line of the springs and the blocks as shown in Fig. 4.2. Let x_1 and x_2 be the displacements of the two blocks from their equilibrium positions. Note that x_1 and x_2 are not the x-coordinates of the blocks but rather the x-components of their displacements from the corresponding equilibrium positions. Let k_{s1} be the spring constant of the spring connecting block m_1 to the left support, k_{s2} the spring constant of the spring connecting block m_2 to the right support, and k_c the spring constant of the spring coupling the two blocks as indicated in the figure.

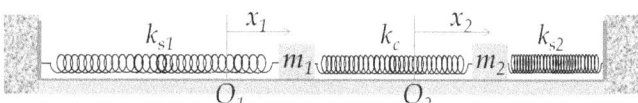

Fig. 4.2: Two blocks coupled with each other by a spring and attached to fixed supports with additional springs. At equilibrium, the blocks are at points marked O_1 and O_2. The x-component of the displacements of the two blocks as measured from O_1 and O_2 are denoted by x_1 and x_2. Any change in x_1 affects the forces on block # 2 and any change on x_2 affects the forces on block # 1. We say that the motions of the blocks are coupled.

The x-components of the equations of motion of the two blocks are

$$m_1 \frac{d^2 x_1}{dt^2} = -k_{s1} x_1 - k_c(x_1 - x_2), \qquad (4.1)$$

$$m_2 \frac{d^2 x_2}{dt^2} = -k_{s2} x_2 + k_c(x_1 - x_2). \qquad (4.2)$$

Rather than solve these two equations for arbitrary values of the masses and constants, let us simplify the problem by looking at the case where masses are equal and two of the springs connecting the blocks to the support have equal spring constants.

$$m_1 = m_2 \equiv m$$

$$k_{s1} = k_{s2} \equiv k_s$$

Then, the equations of motion for x_1 and x_2 are

$$m\frac{d^2x_1}{dt^2} = -k_sx_1 - k_c(x_1 - x_2), \tag{4.3}$$

$$m\frac{d^2x_2}{dt^2} = -k_sx_2 + k_c(x_1 - x_2). \tag{4.4}$$

To solve these equations, we introduce new composite variables X and x in place of x_1 and x_2 by

$$X = \frac{x_1 + x_2}{2}, \tag{4.5}$$

$$x = x_1 - x_2. \tag{4.6}$$

Equations 4.3 and 4.4 then gives rise to the following equations of motion for $X(t)$ and $x(t)$.

$$m\frac{d^2X}{dt^2} = -k_sX, \tag{4.7}$$

$$m\frac{d^2x}{dt^2} = -\left(k_s + 2k_c\right)x. \tag{4.8}$$

Equation 4.7 is an equation of motion of an oscillator of mass m and spring constant k_s, and Eq. 4.8 is an equation of motion of an oscillator of mass m and spring constant $k_s + 2k_c$. Therefore, we can think of X and x as displacements of two fictitious simple harmonic oscillators of mass m attached to springs of spring constants k_s and $(k_s + 2k_c)$ respectively. The solution of these equations show that the variables X and x will oscillate with angular frequencies given by

$$\text{X oscillates at:} \quad \omega_X = \sqrt{\frac{k_s}{m}} \tag{4.9}$$

$$\text{x oscillates at:} \quad \omega_x = \sqrt{\frac{k_s + 2k_c}{m}} \tag{4.10}$$

The variables X and x are called **normal modes** and the angular frequencies ω_X and ω_x the **normal frequencies** of this system of two masses. The normal coordinates X and x execute simple harmonic motions of their corresponding frequencies and are given by

$$X(t) = A\cos(\omega_X t - \delta) \tag{4.11}$$

$$x(t) = B\cos(\omega_x t - \phi) \tag{4.12}$$

The coefficients of x_1 and x_2 in the definitions of X and x in Eqs. 4.5 and 4.6 show how the two masses move when the system oscillates in mode X or x.

Thus, Eq. 4.5 shows that when the masses are moving in the mode X, the two masses have equal displacements and are in-phase as reflected by the same sign for x_1 and x_2 in the expression for X given in Eq.4.5. When the two masses are moving in this mode only the springs on the outside to the supports contract or expand while the

Mode X

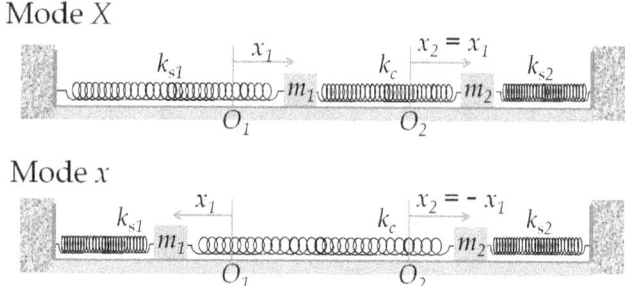

Mode x

Fig. 4.3: Illustrations of the two normal modes X and x. In mode X, the blocks move in tandem, both towards right or both towards left, such that the middle spring in unaffected. In mode x, the blocks move in the opposite directions such that the center of mass of the two blocks remains fixed.

coupling spring remains un-affected as shown in Fig. 4.3. That is why the frequency of this mode does not have dependence on k_c.

On the other hand, Eq. 4.6 shows that when the motion is in the mode x, the displacements of the two masses have equal magnitude but are in opposite directions as reflected in opposite signs for x_1 and x_2 in x given in Eq.4.6. In this mode, the center of mass remains fixed, which can happen if the masses are moving in opposite directions as shown in Fig. 4.3. When the two masses move according to this mode, all three springs participate, which is reflected in the dependence of the frequency of this mode on both k_s and k_c.

4.2 NORMAL MODES OF A STRING

Newton's second law can be applied to oscillations of a string such as guitar string. Consider a string tied at both ends. First, we divide up the mass of the string in N parts and place them as beads on equal intervals as shown in Fig. 4.4. This converts the problem of vibration of a string to the vibrations of N beads, each of mass m, on a massless string which are coupled by the tension force in the string. For the sake of ease of visualization of the modes we consider the modes of the transverse motion of the string.

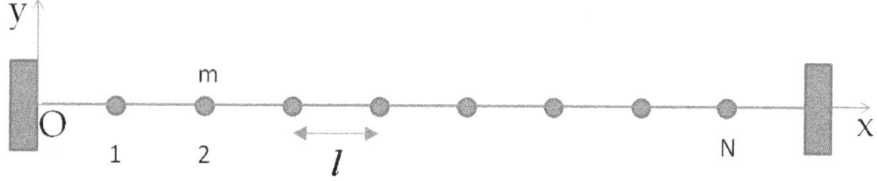

Fig. 4.4: N beads on a taut string act as N coupled oscillators.

There are N normal modes of frequencies for this system.

$$\omega_n = \sqrt{\frac{4T}{ml}}\sin\left[\frac{n\pi}{2(N+1)}\right], \quad n = 1, 2, \cdots, N. \qquad (4.13)$$

where T is the tension in the string, m mass of each bead, and l the separation between beads.

Let the string be along x-axis, then the transverse coordinates of the beads will be in the yz plane. Let us orient y-axis to be the axis in which the transverse motions of the beads occur. With this choice we display the y-displacements of the beads in different normal modes for some particular values of N in Fig. 4.5.

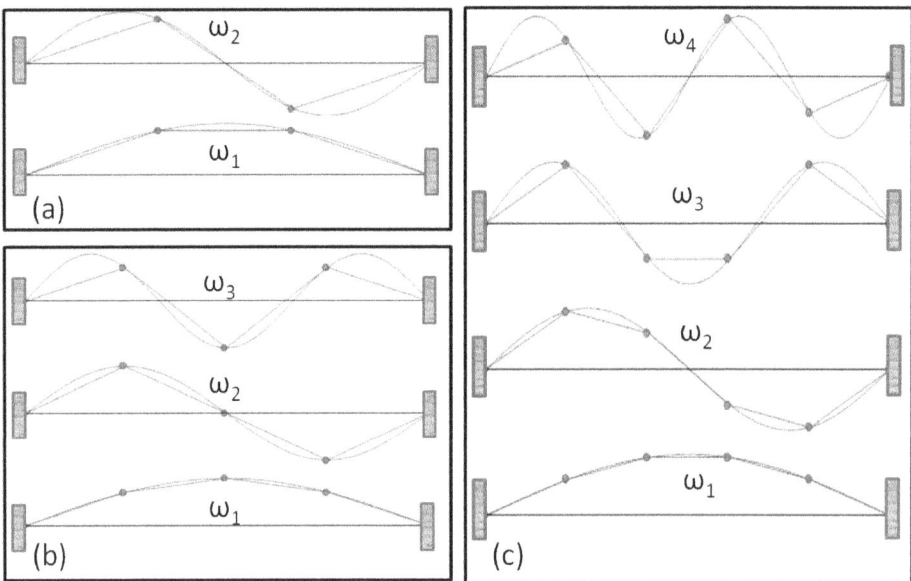

Fig. 4.5: Transverse modes of (a) $N = 2$, (b) $N = 3$, and (c) $N = 4$. In each case ω_1 is the lowest frequency mode. The points at the nodes do not oscillate; other points of the string oscillate up and down with the frequency of the normal mode.

With the origin at one end, the modes can be written as a sine function of x where the x values are x coordinates of the beads, which can be worked out explicitly. For mode of frequency ω_n the pattern along the string of length L that is fixed at both ends is

$$y_n(x) = A_n \sin\left(\frac{\pi x}{L}\right), \quad 0 \le x \le L. \qquad (4.14)$$

The oscillation properties of the continuous string would be obtained by taking $N \to \infty$ limit in Eqs. 4.13 and 4.14, which corresponds to decreasing indefinitely the distance l between elementary masses of the string. By taking this limit we obtain the following normal mode frequencies for a string of length L, mass per unit length μ which is under tension T.

$$\omega_n = n\omega_1, \quad n = 1, 2, 3, \cdots, \tag{4.15}$$

where

$$\omega_1 = \frac{\pi}{L}\sqrt{\frac{T}{\mu}} \tag{4.16}$$

is the frequency of the lowest mode, which is also called the **fundamental mode**, or simply the fundamental. The frequency of the upper modes are integral multiples of the frequency of the fundamental mode. Therefore, the upper modes are also called **overtones**.

4.2.1 Transverse Vibrations of a String

In a transverse vibration mode of a string, each part of the string vibrates up and down with the frequency corresponding to that of the mode. To get a feel for the oscillations of the string, we will plot some of the lowest frequency modes over one complete cycle. Figure 4.6 shows the lowest frequency mode. We find that the lowest frequency mode is symmetric about the middle of the string.

Suppose, we pull the string such that string takes the shape of the lowest mode, and release the string from rest. What will happen? Fig. 4.6 shows the position of the string at intervals of $\frac{1}{8}^{th}$ periods. The string moves slowly in the first $\frac{1}{8}^{th}$ period when it is furthest from the equilibrium than in the second and third $\frac{1}{8}^{th}$ periods when it passes through the equilibrium position, which is similar to the behavior of a mass attached to a spring. The particles near the middle of the string cover the largest distance in the same time than the particles near the fixed ends, and all particles of the string vibrate with the same period of oscillation. Each particle of the string "acts" as if the particle was attached to an "invisible" spring and vibrates up and own in a sinusoidal manner corresponding to the Simple Harmonic Motion at the frequency of the mode.

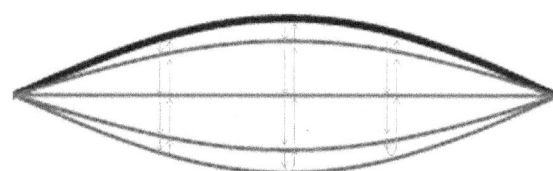

Fig. 4.6: The fundamental mode of a string fixed at both ends. The string is plucked in the shape shown in thick black line and let go from rest. The string's location at successive $\frac{1}{8}$th period intervals are shown in the figure. The string vibrates such that the ratio of the vertical displacements at different horizontal locations remains independent of time.

The first overtone, i.e. mode $n = 2$, is shown in Fig. 4.7. The first overtone mode is antisymmetric about the middle position. When the left half of the string moves up, the right half moves down, and vice-versa with every particle oscillating with frequency ω_2. The point in the middle remains at rest and is called a **node**.

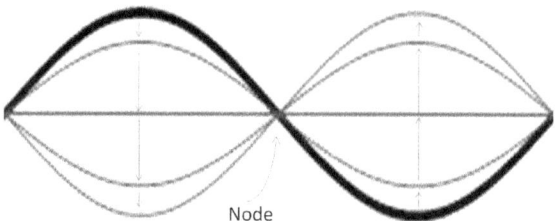

Fig. 4.7: The second lowest frequency mode, $n = 2$, of a string fixed at both ends. The stretched string is pulled in the shape shown in thick black line and let go from rest. The string's location at successive $\frac{1}{8}^{th}$ period intervals are shown in the figure. The point labeled "Node" is stationary and does not move.

4.3 OSCILLATIONS OF A VERY LONG STRING

In the last section we studied the normal modes of a string of length L that was fixed at both ends. We found that the string supports the normal modes in which the string can oscillate at various frequencies that are integral multiples of a fundamental frequency.

Now, we ask: what happens when the string in very long? Clearly, if the string is "infinitely" long, we will be looking at only a small finite part of the string that would be far away from the two ends. Suppose, again, the string to be along the x-axis. If the string is infinitely long, we can assume that the oscillations occur over the entire x-axis. Sinusoidal oscillations that continue over the entire axis are periodic functions, and we write the modes as

$$y(x) = A\sin(kx), \quad -\infty < x < \infty, \qquad (4.17)$$

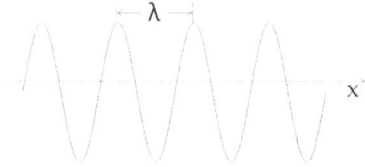

Fig. 4.8: Wavelength corresponding to a mode of oscillation of a taut string.

where k is called the **wavenumber**. For an infinitely long string, the wavenumber can assume any real value up to a maximum cutoff value, which depends on the smallest distance between atoms of the string. The period of the sinusoidal mode function given in Eq. 4.17 is called the **wavelength** corresponding to the mode. Wavelength is usually denoted by the Greek letter λ (read: lambda). From the periodicity of Eq. 4.17 along the x-axis,

$$y(x + \lambda) = y(x), \quad -\infty < x < \infty,$$

we see that the wavelength is related to the wave number by

$$\lambda = \frac{2\pi}{k}. \tag{4.18}$$

4.4 TRAVELING WAVES

A traveling or progressive wave represents the movement of a disturbance from one place to another without the actual movement of the underlying medium itself. The property characterizing a disturbance depends on the nature of the wave. For instance, in the case of a mechanical wave or sound wave, the disturbance refers to the displacement of the particles of the medium from their equilibrium positions as we have seen above. In the case of the electromagnetic wave, the disturbance refers to is the change in the electric and magnetic fields in space. The disturbance varies both in space and time. The function that describes the disturbance is called the **wave function**. We have used $y(x, t$ for the transverse vibration modes on a string. We will use symbol ψ (read: psi) for a wave equation which may stand for a traveling transverse vibrations on a string or some other system.

A particularly simple wave function for the mathematical analysis results if the continuous disturbance can be cast as a sine or a cosine function of the space and time variables. These waves are called **sinusoidal waves** and are characterized by periodicities of the wave in space and time. These waves are also called **plane waves** if they travel in a three-dimensional space since the wave function for these waves take the same value in a plane perpendicular to their propagation direction.

The wave function ψ for a sinusoidal wave of amplitude A, wavenumber k, and angular frequency ω traveling towards the positive x-axis is given by

$$\psi(x,t) = A\,cos(kx - \omega t), \quad -\infty < x < \infty, \ -\infty < t < \infty \tag{4.19}$$

assuming $\psi(0,0) = A$, and k and ω would be taken to be positive. To visualize the motion of a wave we usually draw multiple images of the **wave profile** at successive times.

In Fig. 4.9 I have plotted snapshots of a sinusoidal wave at different times of a plane wave traveling towards the positive x-axis. Note that the wave arrives at a screen at different times with different values

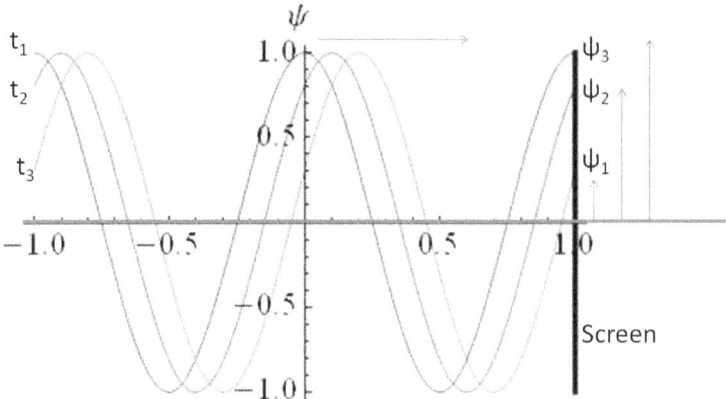

Fig. 4.9: A sinusoidal traveling wave along the x-axis for three different instants $(t_3 > t_2 > t_1)$. The wave function $\psi = cos[2\pi(x - t)]$ is plotted with $t = t_1 = 0$, $t = t_2 = 0.1$ sec, and $t = t_3 = 0.2$ sec. Note that the wave arrives at the screen with different values of wave function at different times.

of the wave function. The picture of the wave function at different times shows that the wave at the screen or at any other point in space acts as a simple harmonic oscillator, oscillating between $\psi = A$ and $\psi = -A$ with frequency ω as the wave passes through the point.

The wave function in Eq. 4.19 moves towards positive x-axis. If the two terms in the argument of cosine in the wave function in Eq. 4.19 have the same sign, the wave would move towards negative x-axis. Therefore, the following wave function will be for a wave moving towards the negative x-axis.

$$\psi'(x, t) = A \; \cos(kx + \omega t). \tag{4.20}$$

At any time, say $t = 0$, the successive crests or successive troughs of the wave are separated by a wavelength, λ, which is $2\pi/k$. You can see this by setting $t = 0$ in either Eq. 4.19 or 4.20 and finding the period of the resulting function, $A\cos(kx)$.

As seen in Fig. 4.9 the wave $A \; cos(kx - \omega t)$ moves towards the positive x-axis. If we focus on one of the crests, we note that the crest moves over a distance equal to a wavelength in time $2\pi/\omega$. Therefore, the speed of the wave v is given by dividing $|\Delta x| = \lambda = 2\pi/k$ by $\Delta t = 2\pi/\omega$.

$$v = \frac{|\Delta x|}{\Delta t} = \frac{\omega}{k} \tag{4.21}$$

We can write the speed of the wave in terms of the wavelength λ and the regular frequency f rather than in terms of the wavenumber and the angular frequency.

$$v = \frac{\omega}{k} = \lambda f. \tag{4.22}$$

4.5 STANDING WAVES ON A TAUT STRING

Each normal mode of a stretched string actually consists of two traveling waves, one moving to the right and the other moving to the left as we will see below. We have determined above that the transverse vibrations of a stretched string fixed at both ends are given by the following displacement function.

$$y_n(x,t) = A_n \sin(k_n x) \cos(\omega_n t - \delta_n), \quad 0 \le x \le L, \qquad (4.23)$$

where the wavenumber and frequencies of the modes are

$$k_n = n\pi/L \ \text{ and } \ \omega_n = n\omega_1,$$

with

$$\omega_1 = \sqrt{\frac{T}{\mu}}\frac{\pi}{L},$$

and $n = 1, 2, 3, \cdots$. Here A_n is an amplitude of the nth mode and δ_n the corresponding phase of the mode. The amplitude and phase depend on the initial conditions on the string. We will set $\delta_n = 0$ to focus on other properties that are insensitive to the initial conditions. The following trig identity helps one write the product of a sine and a cosine as a sum of two sine functions.

$$2\sin A \cos B = \sin(A - B) + \sin(A + B).$$

Using this identity we write Eq. 4.23 as a sum of two sine functions

$$\begin{aligned} y_n(x,t) = \frac{A_n}{2} [&\sin(k_n x - \omega_n t) \\ &+ \sin(k_n x + \omega_n t)], \quad 0 \le x \le L, \qquad (4.24) \end{aligned}$$

The first term in Eq. 4.24 is a wave traveling towards the positive x-axis and the second term is a wave traveling towards the negative x-axis. Therefore, the normal modes of vibration of a string $y_n(x,t)$ turns out to be a sum of two waves, $\sin(k_n x - \omega_n t)$, which moves to the right towards the positive x-axis, and $\sin(k_n x + \omega_n t)$, which moves to the left towards the negative x-axis. Since the two waves have the same wavelength and same frequency, they travel at the same speed v.

$$v = \frac{\omega_n}{|k_n|} = \frac{n\sqrt{\frac{T}{\mu}\frac{\pi}{L}}}{n\pi/L} = \sqrt{\frac{T}{\mu}}.$$

We can imagine the way a standing wave is set-up: the right-moving wave is reflected off from the right boundary, and the left-moving wave is similarly reflected off from the left boundary. The waves then interfere with each other and create a standing pattern, called the standing wave, which shows only the up and down vibrations of each element and no traveling wave.

4.5.1 Forced Oscillations of a String

One way of vibrating a stretched string in one of the normal modes is to pull the string in the shape of the normal mode profile, and then releasing the deformed string from the rest. The subsequent motion of the string will be the oscillatory motion of the mode. Although this method of exciting a string to oscillate in a normal mode is possible, but it is very difficult to achieve.

Oscillating the string at one end with a frequency of the normal mode can also set the string into the mode as illustrated in Fig. 4.10 and 4.11. An oscillating force at one end near the node of a normal mode generates a wave that travels towards the fixed end at the wave speed $v = \sqrt{T/\mu}$, where T is the tension in the string and μ mass per unit length. When the wave arrives at the fixed end, a reflected wave is generated that has the phase opposite to that of the incident wave. The two waves travel in the opposite directions. If the oscillations correspond to one of the normal modes, then the right-traveling and left-traveling waves add such that the sum is a standing wave. A student will benefit by playing with simulation programs of the standing wave generation available online at various sites. One good site is $http : //phet.colorado.edu/sims/wave\text{-}on\text{-}a\text{-}string/$ $wave\text{-}on\text{-}a\text{-}string\_en.html$.

We driving a standing wave by attaching the string to an oscillator near a node of the corresponding normal mode as shown in Fig. **??**. This experimental situation presents conditions on the ends of the string, called **boundary conditions**. We can write the boundary conditions on the wave equation analytically as:

$$y(x,t)|_{x=-\epsilon} = A_0 \, cos(\omega t), \qquad (4.25)$$

$$y(x,t)|_{x=L} = 0, \qquad (4.26)$$

where $\epsilon << \lambda$ and $x = \epsilon$ is near the node at the origin and A_0 is the the amplitude of the oscillation at the oscillator. When the frequency ω is near the frequency of a normal mode, the amplitude of the wave A is much larger than A_0, and the place of the vibrator can be considered to be at a node. The solution of the wave problem with conditions give in Eqs. 4.25 and 4.26 can be written as

$$y(x,t) = Asin(kx)cos(\omega \, t). \qquad (4.27)$$

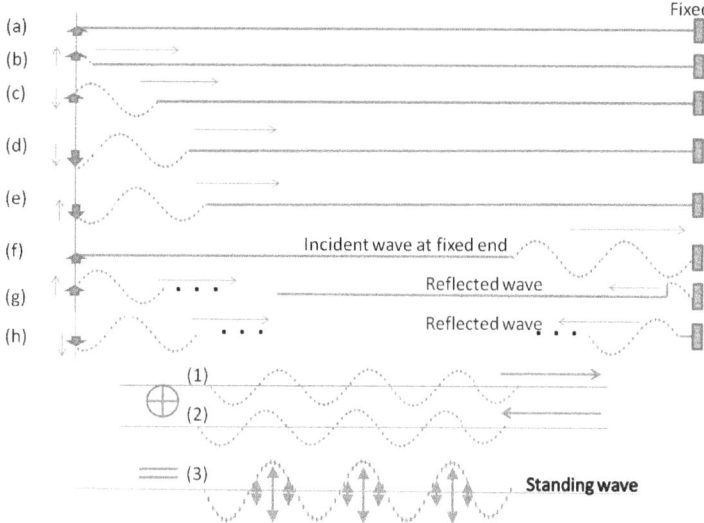

Fig. 4.10: Generation of standing wave by driving vibration at a node of a normal mode. (a) The stretched string at equilibrium. (b)-(e): the right-moving traveling wave generated as the vibrator moves up and down. (f) the incident wave at the fixed support. (g) the generation of the reflected wave with the phase flipped with respect to the incident wave. (h) the reflected wave traveling back. (1)-(3) shows the addition of two traveling waves generates the standing wave in which the particles vibrate in phase with a frequency equal to the frequency of the normal mode.

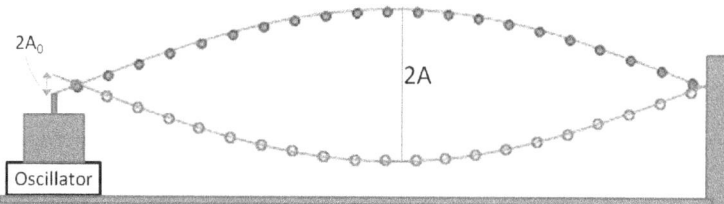

Fig. 4.11: Generation of standing wave by driving vibrations by an oscillator near the node of a normal mode.

This will satisfy Eqs. 4.25 and 4.26 if

$$A \sin(k\epsilon) = -A_0, \tag{4.28}$$

$$\sin(kL) = 0. \tag{4.29}$$

Note that Eq. 4.29 is satisfied automatically for the mode of the string with nodes at $x = 0$ and $x = L$, and Eq. 4.28 gives the relation between the amplitude of the oscillator near the origin and the amplitude of the wave riding on the string. The wave function given in Eq. 4.27 is equal to the sum of two waves - one traveling to the right and the other traveling to the left.

$$y(x,t) = \frac{A}{2} \left[sin\left(kx - \omega t\right) + sin\left(kx + \omega t\right) \right]. \tag{4.30}$$

Fig. 4.11 shows the lowest normal mode driven by an oscillator. A string under tension is tied to the plunger of an oscillator. If the oscillator vibrates at the mode frequency of the lowest frequency mode, we find the string oscillate at a magnified amplitude since the oscillator is near a node confirming the relation between the distance A_0 traveled by the plunger of the oscillator and the amplitude A of the wave.

4.6 POLARIZATION OF WAVES

The displacement from the equilibrium is often a vector property. The relative direction of the displacement and the direction of travel of wave defines the **polarization** of the wave. If the displacements of particles of the medium happen perpendicular to the direction of travel of wave, e.g. when a rope is moved up and down at one end while the wave travels down the rope, we say that the wave is **transversely polarized**. In some waves, e.g., sound waves in air, the particles of air vibrate along the same axis as the wave travel. These types of waves are said to **polarized longitudinally**.

The longitudinal and transverse waves are easily demonstrated by waves on a slinky (Fig. 4.12). Tie one end of a slinky to a post or ask a friend to hold it firmly. Hold the other end so that slinky is a little extended and taut.

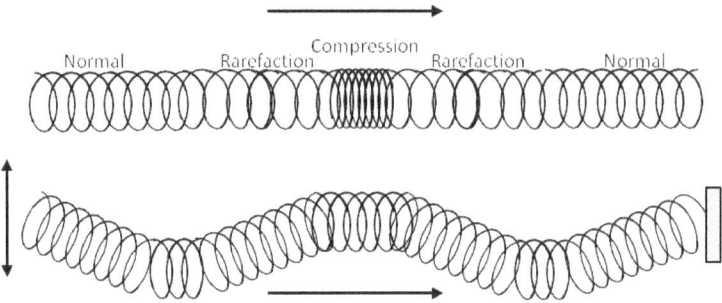

Fig. 4.12: Longitudinal and transverse waves on a slinky.

Now, you can set either a longitudinally polarized or a transversely polarized wave down the length of slinky. To get a longitudinally polarized wave, bunch a few turns together and let go, you will notice that the slinky winds and unwinds keeping the bunch shape of the disturbance more or less intact. To get a transversely polarized wave, pull one of the loops little to the right, left, up or down and let go. The transverse wave will travel down the slinky as it does for the rope described above.

4.7 EXAMPLE WAVE FUNCTIONS

In this section, we will present two examples of wave functions that are very useful for analytic purposes.

4.7.1 Plane Wave

A plane wave is a simple three-dimensional wave where the wave travels along an axis and the wave function has the same value throughout any plane normal to the direction of wave. If the wave is traveling along x-axis, then on each plane parallel to the yz plane, the wave displacement has the same value (see Fig. 4.13). Therefore, the wave function for a plane wave traveling towards the positive x-axis would not depend on the y or z-coordinates of the point. The wave-fronts of such waves are planar as shown in Fig. 4.13.

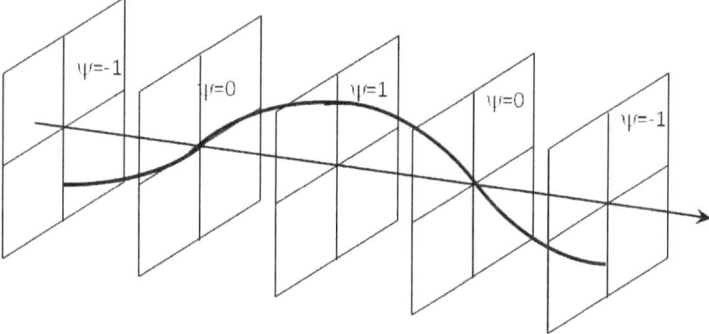

Fig. 4.13: A plane wave has same wave amplitude in the entire plane. It is a volume wave. Although it looks similar to the wave on a chord it is three-dimensional rather than one-dimensional.

The analytic expression of the wave function ψ for a plane wave of amplitude A, wavenumber k, angular frequency ω, traveling towards the positive x-axis, passing through the origin at $t = 0$ with the displacement equal to A is identical to the sinusoidal wave on a string since the wave is essentially one-dimensional.

$$\psi(x,t) = A\,cos(kx - \omega t). \tag{4.31}$$

Similar to the wave function for waves on a string, other possibilities at the origin at time $t = 0$ are taken into account by introducing a phase constant ϕ in the argument of the cosine and sine functions.

$$\psi(x,t) = A\,cos(kx - \omega t + \phi). \tag{4.32}$$

The following more general function describes the displacement for a plane wave traveling in the direction of the wave vector $\vec{k}$.

$$\psi(x, y, z, t) = A\, cos(\vec{k} \cdot \vec{r} - \omega t + \phi), \qquad (4.33)$$

where $\vec{r} = x\hat{x} + y\hat{y} + z\hat{z}$. The magnitude of the wavevector is the wavenumber of the wave, and the wavelength λ of the wave is related to the magnitude of the wavevector as, $\lambda = 2\pi/|\vec{k}|$.

4.7.2 Spherical Wave

If the energy from a source spreads out radially, such as the optical energy from the sun, it is more appropriate to use spherical coordinates to describe these waves as shown in Fig. 4.14.

$$\psi(r, t) = \frac{A}{r}\, cos(kr - \omega t + \phi), \qquad (4.34)$$

where r is the radial distance from the origin.

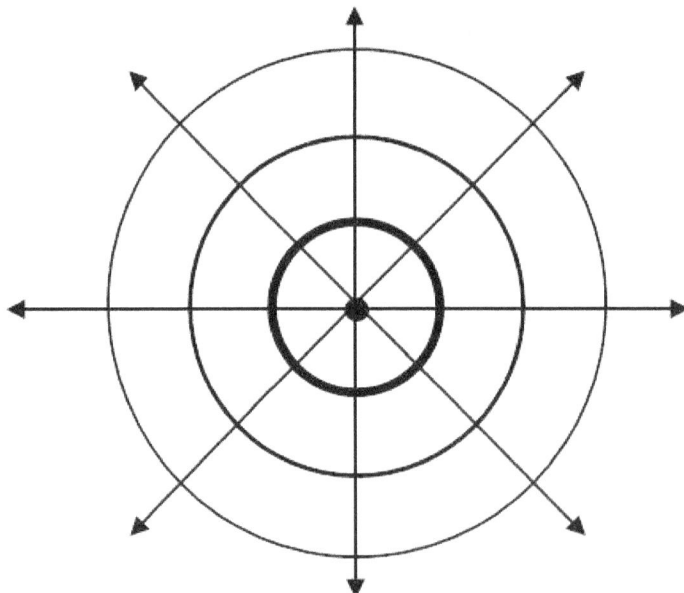

Fig. 4.14: Spherical wave fronts emanating from a source at the center. Same amplitudes at all points that are equal distance from the source at the center. The amplitude drops with the radial distance from the source, shown with drop in thickness of lines, as the distance from the source increases.

4.7.3 Wave Pulses

So far we have discussed waves of infinite extent such as the sinusoidal and plane waves. Sinusoidal waves are given by a sine or a cosine function of space and time. For these types of waves, the displacement oscillates in time according to a combination of sine and cosine functions of time having a definite frequency. To be periodic in space and time, the sinusoidal functions must have a domain that cover the entire space and entire time, viz. $-\infty < x < \infty$ and $-\infty < t < \infty$ for a wave traveling along x-axis. These waves are idealization of the situations encountered in real life.

Real waves, such as water waves, a pulse traveling on a string, or sound of a drum are not sinusoidal waves. They can be of any shape. For instance, suppose a very long taut string is given a snap at one end once and then the end is held fixed in one place as shown in Fig. 4.15. This results in a bump in the string, which moves towards the other end. Similarly, suppose you turn a Laser on for a very brief period, you would get a pulse of light wave. Or, if you hit a piano key, there again you will get a pulse of sound wave. The wave functions for these waves cannot be described by a single wavelength or a single frequency.

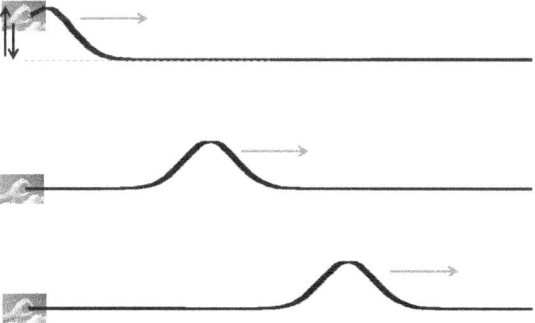

Fig. 4.15: Generation and movement of a wave pulse on a string.

4.8 ENERGY, POWER AND INTENSITY

All waves transport energy without an actual transport of material. For concreteness, consider a wave of frequency f, wavelength λ, and amplitude A on a taut string with tension T and mass per unit length μ. Then, it is possible to prove that the energy E_λ in the wave over the space of one wavelength would be

$$E_\lambda = 2\pi\mu\lambda f^2 A^2 \qquad (4.35)$$

In one period this energy crosses any point of the string. Therefore, the rate of energy flow in the string, i.e. average power in the wave is

$$P_{\text{av}} = \frac{E_\lambda}{1/f} = 2\pi\mu\lambda f^3 A^2 \qquad (4.36)$$

In three-dimensions, wave carry energy through space. For instance, a plane wave $\psi(x, y, z, t) = A\cos(kx - \omega t)$ carries energy towards the positive x-axis. For the waves in the three-dimensional space, we define intensity as the rate at energy passes through planes perpendicular to the direction of the wave per unit area of the planes.

$$\text{Intensity} = \frac{\text{Power}}{\text{Area}} \qquad (4.37)$$

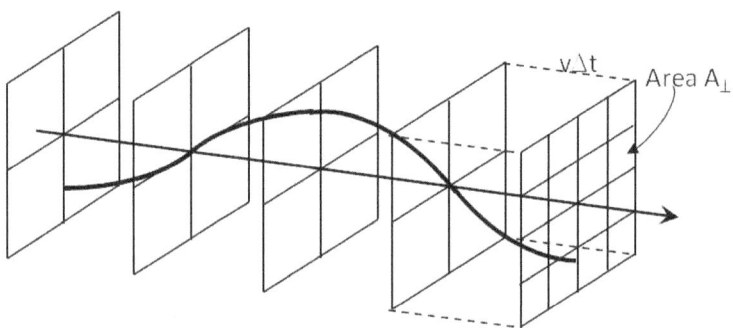

Fig. 4.16: A wave passing through a cross-sectional area; intensity of the wave is defined as average power passing per unit are perpendicular to the wave direction. The wave shown here has speed v. The energy contained in the box of area $A_\perp$ and height $v\Delta t$ pass through $A_\perp$ in time Δt.

Example 4.1. Power of a wave. *The amplitude of a wave on a long string is increased from 10 cm to 20 cm while keeping the same frequency. How is the intensity of the waves changed?*

Solution. Solution Taking the ratio of intensity for the two cases, we find that common factor cancels out, and we obtain the important result that intensity goes as square of the amplitude.

$$\frac{I_2}{I_1} = \frac{A_2^2}{A_1^2} = \frac{(20 \text{ cm})^2}{(10 \text{ cm})^2} = 4.$$

The intensity quadruples when the amplitude is doubled.

4.8.1 Intensity of a Spherical Wave

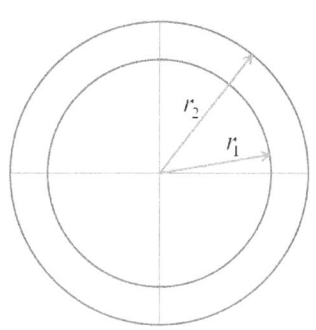

Imagine a point vibrating source that emits a wave isotropically in all directions. As the wave spreads out in three-dimensional space isotropically, i.e., without a preferred direction, the wave will be a spherical wave. The energy passing through the surface at radius r_1 from the source point must equal the energy through the surface at radius r_2.

Therefore, the intensities I_1 and I_2 at two distances r_1 and r_2 respectively from a spherical wave source will follow the following inverse square law.

$$\frac{I_1}{I_2} = \frac{r_2^2}{r_1^2} \quad \text{(spherical wave)} \tag{4.38}$$

Beware that inverse square law for intensity does not hold true if there is a directional preference for the wave or if the wave amplitude does not drop as $1/r$. For instance the amplitude of a plane wave does not drop with distance; hence, the intensity of a plane wave is constant rather than dropping as $1/r^2$ from the source.

$$\frac{I_1}{I_2} = 1 \quad \text{(plane wave)}$$

4.9 WAVE SPEED AND MEDIUM

How fast a mechanical wave travels in a medium depends on the elastic and inertial properties of the medium. If the medium has a high density then it will be more difficult to set the particles of the medium into motion, and therefore we would expect the wave to have slower speed in a more dense medium. On the other hand, if the particles are more tightly bound to each other they will tend to transfer momentum from one to the next more efficiently. Hence we expect the speed of a wave to be larger in stiffer medium.

Dimensionally we can guess the type of relation we might find among the speed of the wave, the density and the stiffness of the medium. The stiffness of a medium is reflected in the bulk modulus B, which is a measure of the fractional change in volume to an applied pressure. Bulk modulus has dimensions of pressure while the density ρ has dimensions of mass over volume. The dimensions of pressure is [Force]/[Area]. Therefore the ratio of bulk modulus to density has dimensions of the square of speed. Hence, dimensionally speaking, speed of a mechanical wave should be proportional to the square root of B/ρ .

$$\text{Dimensions:} \quad [B] \times \frac{1}{[\rho]} = \frac{[L]^2}{[T]^2}$$

That is, we can say that the speed of the wave would be given by

$$v = \alpha \sqrt{\frac{B}{\rho}}$$

where α is an undetermined dimensionless constant. Figuring out the exact relation will require more work, but the dimension analysis gives us some clues about the type of formula we might find.

In a one-dimensional wave on a taut string, the wave speed is related to the tension of the string F_T and the mass density of the string, which is equal to the mass per unit length μ. By applying Newton's second law of motion to a small segment of string it is possible to show that the wave on a taut string has the speed given as

$$v = \frac{F_T}{\mu}, \tag{4.39}$$

which is a ratio of a restorative tendency (tension) and an inertia property (mass per unit length). Similarly, for a wave equation we would obtain speed by replacing F_T by the bulk modulus B and linear density μ by the volume density ρ to obtain the wave velocity in a bulk of a medium

$$v = \frac{B}{\rho}. \tag{4.40}$$

4.10 SUPERPOSITION OF WAVES

When you play two nearby notes on the piano you hear a low frequency beat that was not there before. How could a new note develop from two other notes? Similarly, in a pond when you drop two stones separated by a distance, then at the places the two circular waves meet you find a completely new wave pattern. Is there a way to find out the new wave pattern? Both of these phenomena and many like them result from the creation of the net wave from the sum of other waves overlapping in a region of space at the same time. The principle of the creation of a net wave from the addition of waves is called the **superposition principle**. Suppose there are waves ψ_1, $\psi_2, \cdots$, ψ_N in a region of space at some time t. Then, according to the superposition principle, the waves will superpose to create only one wave that is the sum of all.

$$\psi = \psi_1 + \psi_2 + \cdots + \psi_N. \tag{4.41}$$

The superposition of waves leads to some strange and unexpected consequences as we elaborate now with examples.

4.10.1 Beats

The beat phenomenon of sound is associated with the superposition of two sound waves of slightly different frequencies impinging on the ear drum simultaneously. For simplicity, consider two waves ψ_1 and ψ_2 of different angular frequencies ω_1 and ω_2 and with the same amplitude A meeting at the origin. Suppose, the waves at the origin are represented by the following functions of time.

$$\psi_1(0, t) = A \cos(\omega_1 t)$$
$$\psi_2(0, t) = A \cos(\omega_2 t)$$

In the beat phenomena we are usually interested in two nearby frequencies. Therefore, let us write ω_1 and ω_2 as

$$\omega_1 = \omega_0 - \frac{\Delta\omega}{2}$$
$$\omega_2 = \omega_0 + \frac{\Delta\omega}{2}$$

so that ω_0 is their average frequency and $\Delta\omega$ is the difference between the frequencies. By the superposition principle, the net wave at the origin will be

$$\psi = \psi_1 + \psi_2 = A \left[\cos(\omega_1 t) + \cos(\omega_2 t)\right].$$

Now, the trigonometric identity

$$\cos A + \cos B = 2 \cos\left(\frac{A + B}{2}\right) \cos\left(\frac{A - B}{2}\right)$$

can be used to rewrite this relation as

$$\psi = 2A \cos\left[\left(\frac{\omega_1 + \omega_2}{2}\right) t\right] \cos\left[\left(\frac{\omega_1 - \omega_2}{2}\right) t\right]$$

Let us write this wave function in the terms of the average frequency ω_0 and the difference $\Delta\omega$ of the frequencies.

$$\psi = 2A \cos\left(\omega_0 t\right) \cos\left(\frac{\Delta\omega}{2} t\right) \tag{4.42}$$

The net vibration consists of fast oscillations at the average frequency ω_0 modulated by a slowly varying cosine as shown in Fig. 4.17. The

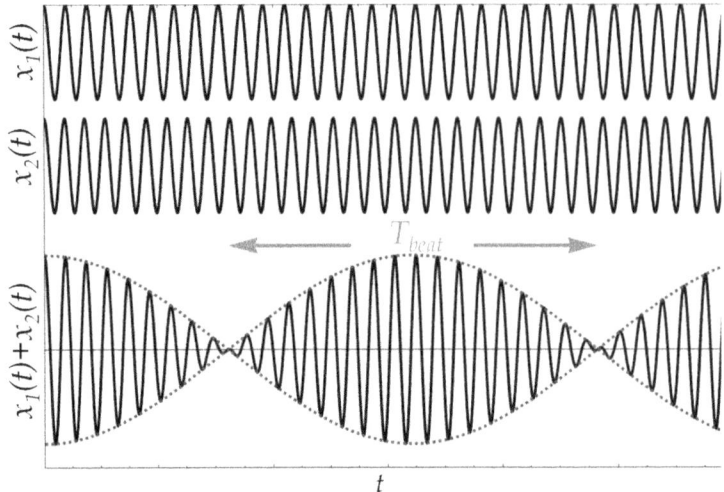

Fig. 4.17: The beats as a result of superposition of two vibrations. The plot was made with $f_1 = 261.63$ Hz and $f_2 = 277.18$ Hz, corresponding to C_4 and $C_4^{\#}/D_4^{b}$ musical notes. The upper graph is the signal at frequency f_1, the middle at frequency f_2 and the bottom graph is the sum of the two. The peaks and troughs of the beats are separated by one beat period as shown.

amplitude of the fast oscillations is not constant in time but rather varies between $-2A$ and $2A$.

The largest amplitude oscillations occur at the peak and trough of the slowly varying modulation $cos\left(\frac{\Delta\omega}{2}t\right)$. If the working frequencies are in the audible range, then the peaks and troughs of the slowly varying modulation corresponds to the loudest sound while the sound is faintest when the modulation function is smallest. Thus, sound from two sources of nearby frequencies appear to go loud and soft periodically - the loud sounds are called beats.

Since both the troughs and peaks of $cos\left(\frac{\Delta\omega}{2}t\right)$ correspond to the loudest sound, there are two beats in one period of this function. Therefore, the beat period is the half of the period of the modulating function $cos\left(\frac{\Delta\omega}{2}t\right)$.

$$T_{beat} = \frac{1}{2}\left(\frac{2\pi}{\Delta\omega/2}\right) = \frac{2\pi}{\Delta\omega}. \tag{4.43}$$

Therefore, the beat frequency is the difference in frequencies of the two component vibrations:

$$\omega_{beat} = \Delta\omega = |\omega_1 - \omega_2|. \tag{4.44}$$

By setting $\omega = 2\pi f$, we can write this relation for the beat frequency.

$$\text{Beat frequency:} \quad f_{\text{beat}} = |f_1 - f_2|. \tag{4.45}$$

4.10.2 Interference of Waves

In the last section we described how waves add to create new waves that may not look anything like the original waves. This is clearly visible when two circular water waves on the surface of a pond overlap. The term **interference** refers to the phenomenon observed in the resulting intensity when the two waves overlap as shown below.

Let ψ_1 and ψ_2 be the wave functions of two waves overlapping in a particular region of space at time t. The resultant wave function would be the sum of ψ_1 and ψ_2.

$$\psi = \psi_1 + \psi_2 \tag{4.46}$$

The definition of intensity of waves says that the intensity is proportional to the time average of the square of the wave function. This gives the following for the resultant intensity where the two waves overlap.

$$I = \alpha < (\psi_1 + \psi_2)^2 > = I_1 + I_2 + I_{12} \tag{4.47}$$

where α is a constant prefactor that relates amplitudes to intensities in the medium, the symbol $< \cdots >$ stands for the time averaging, I_1 and I_2 are the individual intensities of the two waves when the other wave is not present there, and the cross-term I_{12} is called the interference term. Hence, intensity of the sum of the two waves is not equal to the sum of the intensities.

$$I \neq I_1 + I_2 \tag{4.48}$$

While the intensities of the two waves I_1 and I_2 are positive numbers, the interference term I_{12} can be positive or negative. Therefore the actual intensity observed can be different than simply the sum of the intensities of the separate waves. As a matter of fact, if the two waves have equal intensities to begin with, then the net wave can have intensity anywhere between zero and four times the intensity of one of the waves, not just twice.

Mixing two waves of equal amplitude:

For a simple example, we consider two waves of equal frequency and equal amplitude. Let the two waves start out at S_1 and S_2 in synchronization with each other, but wave 1 travels a distance L_1 to reach a point P at the detector while wave 2 travels a distance L_2 as shown in the Fig. 4.18.

This experiment can be set up, for example, by having two slits in a plate through which waves must exit on the other side and move to P where they mix. For instance, we can shine light from a laser on an

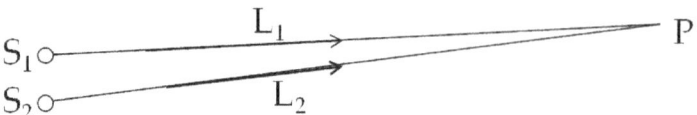

Fig. 4.18: Interference of two waves.

opaque plate with two slits at S_1 and S_2. The experimental arrangement with two slits is also called **Young's double slit experiment** who studied interference of light waves by this method.

Even if the two waves start out in sync at S_1 and S_2, they might not be in sync at point P on the screen, depending upon relative distances from the source points to P. If the difference between L_1 and L_2 is a multiple of a wavelength, then we expect them to end up in sync again at P and a maximum intensity should result there. When this happens, we say that the two waves have a constructive interference at point P. This is shown in Fig. 4.19 where two waves which start out at points S_1 and S_2 in sync end up in sync at point P.

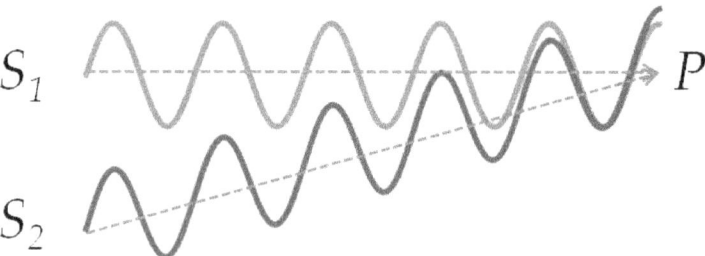

Fig. 4.19: Constructive interference of two waves.

On the other hand, if the difference of distances from the sources at S_1 and S_2 to the point P is an odd multiple of half a wavelength, then the crest of one wave will fall on the trough of the other, making the net amplitude zero there. The waves are said to interfere destructively at point P. If the amplitudes of the waves from S_1 and S_2 are equal, then, the net amplitude at the point of destructive interference will be zero since the two waves will be completely out of sync there. This is shown in Fig. 4.20 where two waves which start out in sync at points S_1 and S_2 end up out of sync at point P and interfere destructively there.

Let us now see how the constructive and destructive interferences show up analytically as a result of the addition of the amplitudes of the two waves. For this, we need to work out the interference term more fully.

$$I_{12} = 2\alpha < \psi_1 \psi_2 >$$

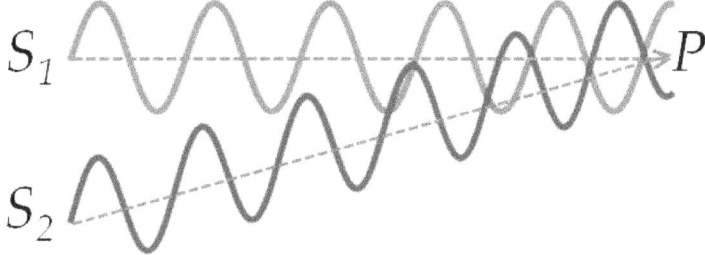

Fig. 4.20: Destructive interference of two waves.

Since waves 1 and 2 travel distances L_1 and L_2 their wave functions at point P will take on the following values at an arbitrary time t.

$$\psi_1(t) = A \cos(kL_1 - \omega t) \tag{4.49}$$

$$\psi_2(t) = A \cos(kL_2 - \omega t) \tag{4.50}$$

Note the waves have the same frequency, wavenumber and amplitude for our exercise here. Note also that we have dropped the phase constant since we have assumed that the waves start out in sync at S_1 and S_2. Multiplying the two wave functions we find

$$\psi_1(t)\psi_2(t) = A^2 \cos(kL_1)\cos(kL_2)\cos^2\omega t + A^2 \sin(kL_1)\sin(kL_2)\sin^2\omega t$$
$$-A^2\left[\cos(kL_1)\sin(kL_2) + \sin(kL_1)\cos(kL_2)\right]\cos\omega t \sin\omega t$$

Since the wave functions are periodic in time, the time average of the product $< \psi_1(t)\psi_2(t) >$ can be done by integrating the product with respect to time over one cycle and then dividing the result by the period.

Time average of a periodic function: $< f(t) >= \dfrac{1}{T}\displaystyle\int_0^T f(t)dt.$

Here the period of the wave function is $T = 2\pi/\omega$. The integration involves separate integrals over $\cos^2(\omega t)$, $\sin^2(\omega t)$, and $\cos(\omega t)\sin(\omega t)$ with the following results.

$$< \cos^2(\omega t) >= \frac{1}{T}\int_0^T \cos^2(\omega t)dt = \frac{1}{2} \tag{4.51}$$

$$< \sin^2(\omega t) >= \frac{1}{T}\int_0^T \sin^2(\omega t)dt = \frac{1}{2} \tag{4.52}$$

$$< \cos(\omega t)\sin(\omega t) >= \frac{1}{T}\int_0^T \cos(\omega t)\sin(\omega t)dt = 0. \tag{4.53}$$

Therefore, the interference term evaluates to

$$I_{12} = \alpha A^2 \cos\left[k\left(L_1 - L_2\right)\right]$$

Note the intensity of each wave is

$$I_1 = \alpha < \psi_1^2 > = \frac{1}{2}\alpha A^2 = I_2$$

Let us write the intensity of the individual waves as I_0.

$$I_0 \equiv I_1 = I_2.$$

Now, writing the intensity observed at point P, all in terms of I_0, the intensity of separate waves, is

$$I = 2I_0 \left\{1 + \cos\left[k\left(L_1 - L_2\right)\right]\right\} \tag{4.54}$$

For fixed source points S_1 and S_2, the intensity depends upon the location of the observation point P since depending upon the value of $k(L_1 - L_2)$ the cosine function fluctuates between -1 and 1. When cosine is equal to -1, the intensity at P will be zero, and when cosine is $+1$, the intensity is four times as much as the intensity in each wave. Hence, we have the following conditions for constructive and destructive interference.

$$k\left(L_1 - L_2\right) = \begin{cases} 2n\pi & n = 0, \pm 1, \pm 2, \cdots \text{ Constructive} \\ (2n+1)\pi & n = 0, \pm 1, \pm 2, \cdots \text{ Destructive} \end{cases} \tag{4.55}$$

Let us write the wave number k in terms of the wavelength to relate to the qualitative pictures we obtained above.

$$\left(L_1 - L_2\right) = \begin{cases} n\lambda, & n = 0, \pm 1, \pm 2, \cdots \quad \text{Constructive} \\ \left(n+\frac{1}{2}\right)\lambda, & n = 0, \pm 1, \pm 2, \cdots \quad \text{Destructive} \end{cases} \tag{4.56}$$

Therefore, the waves from two synchronized sources of the same frequency interfere constructively when the distances to an observation point P differ by multiples of a wavelength and interfere destructively if the distances differ by half a wavelength, or odd multiples of half a wave length.

SIMPLIFICATION FOR LARGE L

Often we are interested in interference of two waves far away from the sources. In this case, we can use trigonometry to simplify the interference conditions when the difference in the path lengths of the two sources to the detector is small compared to the distance to the detector.

From the geometry in the figure, the difference in paths is related to the angle and the separation of the sources.

$$|L_1 - L_2| = d\sin\theta \tag{4.57}$$

Therefore, the conditions for interference are:

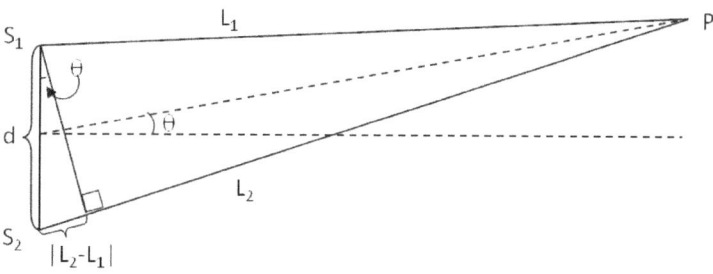

Fig. 4.21: Geometry for approximating the difference in the path length from the source points and a point on the screen.

$$d \sin \theta = \begin{cases} n\lambda, & n = 0, \pm 1, \pm 2, \cdots & \text{Constructive} \\ \left(n + \frac{1}{2}\right)\lambda, & n = 0, \pm 1, \pm 2, \cdots & \text{Destructive} \end{cases} \quad (4.58)$$

Example 4.2. Constructive and Destructive Interference. *Waves of wavelength 50 cm are emitted from two point sources separated by 70 cm. In what directions will there be constructive interferences at a far away detector?*

Solution. This example illustrates the use of the formulas for constructive and destructive interferences in a double-slit experiment. We have been given the wavelength and the distance between the slits. Since the detector is far away, we can use the approximate formula for the interference conditions which gives the directions in which constructive and destructive interferences will occur in terms of the wavelength λ and the distance d between the slits. The constructive interferences are in the directions given by:

$$\sin \theta = \frac{n\lambda}{d}, \quad n = 0, \pm 1, \pm 2, \cdots$$

Clearly, the conditions fails for n for which $\sin \theta > 1$ since sine of an angle must be between -1 and 1, i.e. $-1 \leq \sin \theta \leq 1$. Also, the allowed angles are symmetric about $\theta = 0$ since if $n = 1$ is allowed, then $n = -1$ is also allowed. Now, we systematically work out the allowed angle starting with $n = 0$. We find the following allowed the directions of constructive interference:

$$n = 0 : \quad \theta = 0$$
$$n = \pm 1 : \quad \theta = \arcsin\left(\frac{\pm 1 \times 50\text{cm}}{70\text{cm}}\right) = \pm 45.6°$$
$$n = \pm 2 : \quad \theta = \arcsin\left(\frac{\pm 2 \times 50\text{cm}}{70\text{cm}}\right) \implies \text{no solution}$$

Hence, only three constructive places are possible for the given conditions.

4.11 SOUND

Sound is a mechanical wave responsible for the sensory experience of hearing. We can hear the mechanical waves in air that have frequencies between 20 Hz and 20 kHz, called the audible range. The mechanical waves of frequencies above 20 kHz are called **ultrasound**, and those below 20 Hz the **infrasound**. Sound waves are created by vibrating objects, such as a vocal chord, a guitar string, a drum, etc., and propagate through a medium, most commonly air, from one location to another. Since sound is a mechanical wave, it has all the properties of mechanical waves we have described in the this chapter.

4.11.1 General Properties of Sound Waves

As pointed out above, sound waves are a type of mechanical waves. But unlike waves on a rope, sound waves in air are **longitudinal**, which means that the particles of the medium (air) vibrate along the same line as that of wave travel (Fig. 4.22). As a matter of fact sound wave in all fluids are longitudinal since fluids cannot provide restoring force to a shear stress generated when a sound wave traveling in the medium. Sound waves in solids can be both longitudinal and transverse.

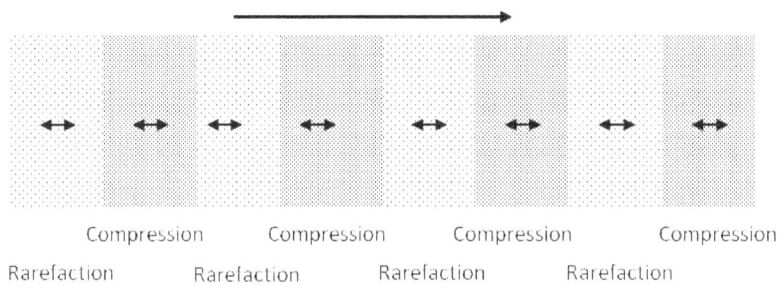

Fig. 4.22: As sound passes in air, it creates regions of high pressure, called compressions, and regions of low pressure, called rarefactions.

When sound travels through air, particles of air vibrating in the forward direction press against other particles of air creating a higher pressure than when the wave was not there. Thus, as sound wave propagates through air we obtain regions of compressed air and rarified air. The phenomena are called **compression** and **rarefaction** of air. Hence, the difference of pressure from the ambient pressure gives us an equivalent description of the disturbance or the wave function.

A simple planar wave can be generated by vibrating a planar surface back and forth sinusoidally. The wave function for the pressure wave will be denoted by Δp to distinguish it from the wave function ψ which will be used for the particle displacement wave. For the sake of continuity in our discussion we postpone deriving the relation between the two equivalent representations of the wave function for sound. The plane wave is a cosine or sine function of position and time as discussed in the last chapter.

$$\Delta p(x,t) = p(x,t) - p_0 = A\cos{(kx - \omega t + \phi_p)} \quad \text{(Planar sound wave)}$$
$$(4.59)$$

Here A is the amplitude of the wave, k the wave number, ω the angular frequency, and ϕ_p the phase constant for the pressure wave. The amplitude A is the maximum deviation of absolute pressure p at a point with the coordinate x at time t from the ambient pressure p_0. The actual frequency f is related to the angular frequency ω as

$$\omega = 2\pi f.$$

The wave number k is related to the wavelength λ inversely as it gives the number of waves that can fit in 2π meters if the wavelength is expressed in meters.

$$\lambda = \frac{2\pi}{k}.$$

The speed v of the plane sound wave comes from the ratio of the distance traveled over time taken, or wavelength divided by one time period T, which is the same as wavelength multiplied by frequency.

$$v = \frac{\lambda}{T} = \lambda f.$$

By analyzing the vibration of the particles of the medium, we can show that the speed of mechanical waves, including sound wave, through a medium comes from a competition between two opposite tendencies - a restoring force whose tendency is to bring the particle to equilibrium and an inertia whose tendency is to maintain the motion. In a one-dimensional system such as a string, the restoring force is provided by the tension in the string (F_T) and the inertia is provided by mass per unit length of the string (μ). The speed of mechanical wave in a string was stated in the last chapter to be

$$v = \sqrt{\frac{F_T}{\mu}}.$$

The speed of sound in air is similarly related to the properties of air. The restoring force is provided by the bulk modulus B and the inertia is provided by mass per unit volume ρ. The speed of sound in air is therefore given in terms of properties of air by the following.

$$v = \sqrt{\frac{B}{\rho}}.$$

The density of air is not constant. It depends on the temperature and pressure. We quote here experimental relation of dependence of speed on temperature. At 1 atm and $0°C$, the speed of sound in air is found to be 331 m/s and at another temperature $t°$C, the speed of sound in air at 1 atm is given by the following approximate formula.

$$v \approx 331 \left(1 + 1.8 \times 10^{-3}t\right) \text{ m/s}.$$

Thus at room temperature of 20°C, the speed of sound in air is approximately 343 m/s. Speed of sound is different in different materials depending upon their bulk moduli and densities. Table 4.1 gives the speed of sound in some common materials of interest.

Table 4.1: Speed of sound in various material at 25°C and 1 atm Source: Kaye & Laby, Table of physical and chemical constants 16th edition (published 1995).

Medium	Speed of longitudinal wave (m/s)	Speed of transverse wave (m/s)
Isotropic Solids		
Aluminum, rolled	6374	3111
Brass	4372	2100
Polycarbonate	2220	910
Pyrex Glass	5640	3280
Steel, Stainless	5980	3297
Liquids		
Blood	1584	
Glycerin (Glycol)	1920	
Mercury	1449	
Water, Distilled	1496	
Water, Sea (3.5% salinity)	1534	
Gases		
Air	343	
Helium $(0°C)$	972	
Hydrogen $(0°C)$	1286	

Most waves encountered in life are not plane waves. For example sound coming from a conical speaker is not plane but more like spherical (Fig. 4.23). At large distances from a conical speaker the curvature of the wave fronts become small enough that the waves can be approximated to be planar.

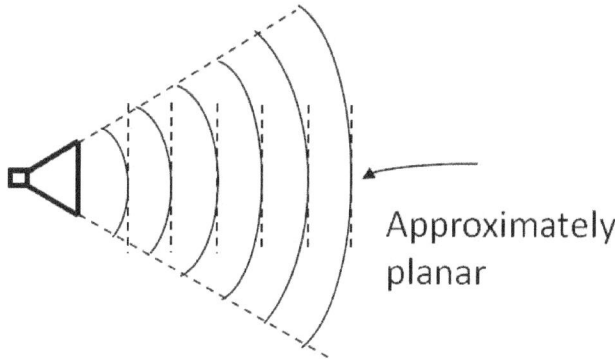

Fig. 4.23: Conical speakers. Far away from the speaker, the waves become almost planar and could be approximated by plane waves.

4.11.2 Intensity of Sound - Decibel

The SI-unit of intensity is W/m^2. The intensity of sound that a human ear can hear ranges from approximately 10^{-12} W/m^2 to 1 W/m^2. This is a very wide range of values. When values of interest are spread over a large range, we use logarithm to collapse the values to smaller range of values. In the case of sound, we introduce a logarithm scale called **decibel(dB)** to capture the exponent of 10 in the absolute intensity I compared to a reference intensity I_0 by base-10 logarithm.

$$I(\text{in dB}) \equiv 10 \log \left[\frac{I}{I_0} \right] \qquad (4.60)$$

Thus if the intensity is 1,000-times the reference, then it will have a decibel of 30, and if it is 1,000,000-times the reference then it has a decibel of 60. The reference intensity for sound wave is 10^{-12} W/m^2, approximately the threshold of normal human hearing at 1000 Hz of frequency. Therefore, human hearing covers 0 to 120 decibels! A useful information is that a 3-dB drop intensity corresponds to halving the signal.

Since absolute intensity is proportional to the square of the amplitude of the wave, the intensity in dB can be also written in terms of amplitude A of the wave to the amplitude A_0 of the reference wave, but this time the factor would be 20 since replacing $I \sim A^2$ leads to a factor of 2 from the logarithm using $log\,(a^n) = n\,log\,a$.

$$I(\text{in dB}) \equiv 20 \log \left[\frac{A}{A_0} \right]. \qquad (4.61)$$

4.11.3 Sources and Quality of Sound

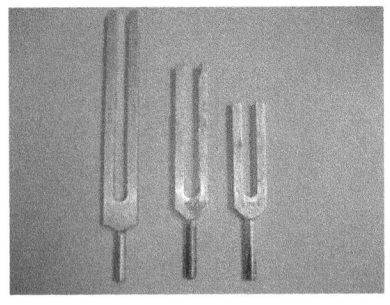

Fig. 4.24: Tuning forks of frequencies 256 Hz, 512 Hz and 1024 Hz, from left to right.

Any vibrating object will be a source of sound. If an object vibrates at a single frequency, it will produce a pure tone of that frequency. For instance, a tuning fork can be constructed that produces a single-frequency sound when struck. Smaller tuning forks produce higher frequencies than the larger ones (Fig. 4.24).

Musical instruments produce discrete set of frequencies, the fundamental and overtones, whose frequencies are integral multiples of the fundamental frequency. When a particular key is played on an instrument, it generally produces both the fundamental frequency and the overtones. You may have noticed that the same key, for instance the middle C of piano, played on different instruments, sound different to our ears. The physical reason for the difference is the different mixture of overtones and their relative amplitudes produced in each. Different mixtures of overtones give the sound its **quality or tone color or timber**.

Not every sound produced are like those of musical instruments, such as when you drop a stone or a rubber ball on a concrete floor. Clearly you can tell whether a stone or a rubber ball hit the concrete floor due to the difference in quality of the two sound, but you cannot discern a pitch in either. Such sounds are called noise. The reason you

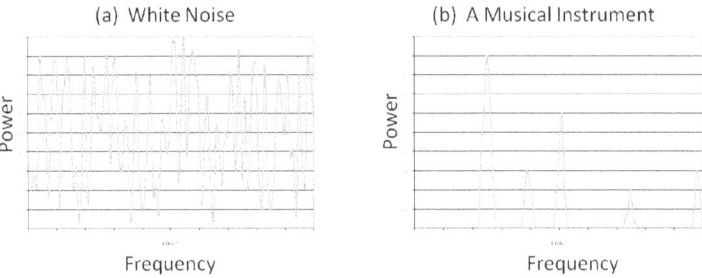

Fig. 4.25: (Power spectra of (a) white noise and (b) a musical instrument. Musical instruments produce fundamental and overtones while noise contains almost continuum of frequencies.

are not able to find a pitch in a noise is that it usually consists of a continuous sound spectrum unlike sound of a musical note which consists of a particular frequency and its overtones. If you plot the intensity versus frequency in a noise, you will get a continuous graph, rather than spikes at the fundamental and overtones in the case of musical notes (Fig. 4.25). The plot of the intensity versus frequency is also called the **power spectrum**.

Perception of sound

Human ears serve as excellent detectors of sound. We perceive several characteristics of sound: loudness, pitch, tone, timbre or tone color. They all have their basis in physical content of the waves impinging upon the eardrums. Loudness has to do with the energy in the wave and is related to the intensity of the wave. Pitch and tone are related with the frequency; a higher pitch or tone generally refers to higher frequency. Timbre or tone color refers to the frequency composition of a musical tone.

4.11.4 Sound through Solid Media

Since sound is a mechanical vibration, it can travel through any material medium. In liquid and gas, only longitudinal sound is possible because fluids do not have a restoring force in tangential direction, they have restoring force only against compression. Solids have restoring force for compression as well as shear forces. Therefore you will find three polarizations of sound waves in solid: one longitudinal, and two transverse modes, one for each perpendicular direction.

The speeds of longitudinal and transverse waves are different since the restoring forces are different for them.

$$v_{\text{sound}} = \sqrt{\frac{\text{Restoring force per unit area}}{\text{Inertia as given by density}}} \qquad (4.62)$$

For a uniform isotropic material the two transverse waves have the same speed. Let Y be the Young's modulus, G the shear modulus, and ρ the density of the solid. The speed of longitudinal sound v_L is related to the Elastic modulus E.

$$v_L = \sqrt{\frac{E}{\rho}} \qquad (4.63)$$

where $E = Y + \frac{3}{4}B$. Here Y is the Young's modulus and B the bulk modulus. On the other hand, the speed of transverse waves v_T is related to the shear modulus.

$$v_T = \sqrt{\frac{G}{\rho}} \qquad (4.64)$$

For instance, steel has $Y \approx 215$ GPa, $B = 166$ MPa, $G \approx 84$ GPa, and density $7,800$ kg/m^3, therefore the longitudinal and transverse waves travel at different speeds in steel.

In steel: $v_L = 6592$ m/s; $v_T = 3281$ m/s.

The numbers here are a little different than those listed in the table because of the temperature dependence of sound. In general, shear stress G is less than the Young's modulus Y. Hence speed of transverse waves will be less than that of longitudinal wave. Sound waves in solids are used to find defects inside solid materials by non-destructive means. The non-destructive techniques based on propagation of waves in material media have important applications in medical physics and other engineering fields. For instance, in aeronautics, invisible cracks in the wings of air planes can be detected even before they become large enough to cause an accident.

4.11.5 Ultrasound

Ultrasound is the name given to the sound waves having frequencies above 20,000 Hz. Humans can't hear these waves, but many animals such as dogs and bats have sensitivity to these waves. Bats even use ultrasound for navigation at night by producing ultrasound that bounces off obstacles in their flight path. In tissue and water, ultrasound of frequency 5 MHz range have wavelength in the millimeter and sub-millimeter range. Since a wave scatters most strongly from obstacles that have dimensions of the order of their wavelengths, ultrasound sent inside human body is able to reflect off inter-tissue spaces of millimeter dimensions. The returned wave is constructed into image of the reflecting surfaces giving ultrasound imaging of internal organs. Ultrasound machines have become universally accepted as powerful tools for the detection of kidney stones, heart ailments and fetuses. Powerful ultrasound machines are also used for surgery.

Ultrasound also has many industrial uses, including cleaning, and non-destructive technology where one can detect defects in inside metals sheets that otherwise appear normal.

4.12 DOPPLER EFFECT

So far in our discussion, there was no motion of detector or source of wave with respect to the medium. Hence, in all the foregoing studies, we have assumed that the frequency of wave detected is same as that of the source. However, if either the source or the detector moves with respect to the medium, there will be a difference in the frequency detected at the detector and that produced at the source. The effect is called the **Doppler effect**, named after the Austrian physicist

Christian Johann Doppler (1803-1853), who in 1842 proposed the effect in his book titled, "Uber das farbige Lict der Dopperlterne," which means "the colored light of double stars".

Doppler effect also provides explanation for the high pitch of an incoming train whistle and low pitch of a train moving away. To be sure, the incoming train whistle also appears louder since the whistle is blown in the direction of the platform, but here we are not concerned with the loudness, but the pitch or the frequency. This is also the reason why the spectrum of light is shifted to the lower frequency, the so-called red-shift, if a star or a galaxy is moving away.

To understand the Doppler effect we will first consider two instances, one in which the source is stationary with respect to the medium while the detector moves at a constant velocity, and the other in which the detector is stationary with respect to the medium and the source moves with a constant velocity. Then we will combine the results to figure out what would happen when both the source and the detector move with respect to the medium. To keep it simple the relative motion will be kept along the line joining the source and the detector, which will be taken to coincide with the x-axis with positive x-axis pointed from detector to source.

Moving Detector and Stationary Source

Let v_D be the speed of the detector while the source is fixed with respect to the medium, and v be the speed of the wave in the still medium (Fig. 4.26). Note that if the detector is moving away from

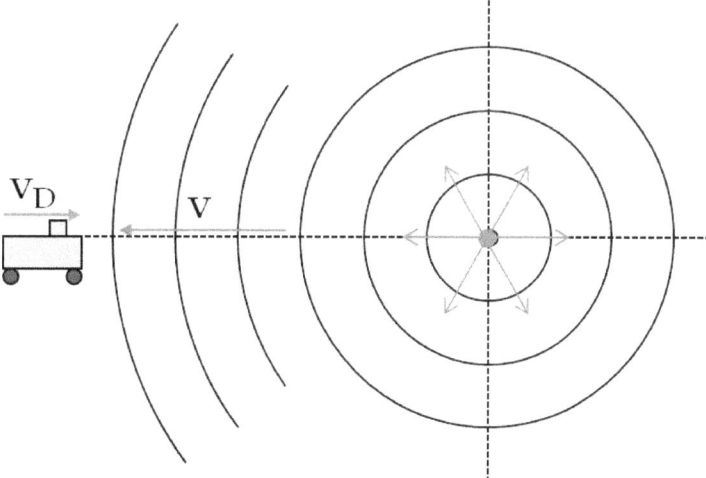

Fig. 4.26: Doppler effect when the detector moves while the source remains stationary. When the detector is moving towards the stationary source, the detector encounters the wavefronts more frequency and therefore it will register higher frequency for the wave. If the detector were to move away from the source, the wave fronts will reach to the detector at longer time intervals and hence the detector will register a smaller frequency.

the source with a speed greater than the speed v of the wave in the medium, then the wave will never catch up. Therefore, we will impose the restriction here that if the detector is moving away from the source its speed be less than the wave velocity.

$$v_D < v \text{ if detector moving away.}$$

We do not need to put any restriction if the detector is moving towards the source. Let f_0 be the frequency of the wave emitted by the source. That is, there will be a new wave front every $1/f_0$ second. Wavefronts from the source will travel towards the detector at speed v with the distance between the wavefronts equal to $\lambda = v/f_0$, which is the wavelength λ.

For nonrelativistic speeds, e.g. v and $v_D << c$, the speed of light in vacuum, the speed of the detector with respect to the moving wave front will be $v + v_D$ if moving toward the source, and $v - v_D$ if moving away from the source. Hence, the distance λ between the wavefronts will be covered by the detector in different time then $1/f_0$. Let T_D be the time the detector encounters the successive wavefronts. Then T_D will be

$$T_D = \frac{\lambda}{v \pm v_D} \begin{cases} + & \text{when detector moves towards the source} \\ - & \text{when detector moves away from the source} \end{cases}$$
$$(4.65)$$

The frequency of the wave detected by the detector will be the inverse of this time period. Therefore, the frequency detected by the moving detector will be

$$f = \frac{1}{T_D} = \left(\frac{v \pm v_D}{v}\right) f_0 \begin{cases} + & \text{when detector moves towards the source} \\ - & \text{when detector moves away from the source} \end{cases}$$
$$(4.66)$$

Thus, when the detector moves towards the stationary source, the frequency at the detector is higher than that produced by the source, and when the detector is moving away from the stationary source, the frequency detected is lower than the one produced. This explains why the pitch appears higher if you run towards a stationary horn and lower if you run away from the stationary horn.

Moving Source and Stationary Detector

Consider a source of wave that moves with a speed v_S with respect to the medium in which the wave travels at the speed v while the detector is fixed in its place with respect to the medium as shown in Fig. 4.27.

Even though the source is moving, it still emits wave at the same frequency f_0. Waves move out in expanding spheres centered at the the instantaneous location of the source which is changing in time

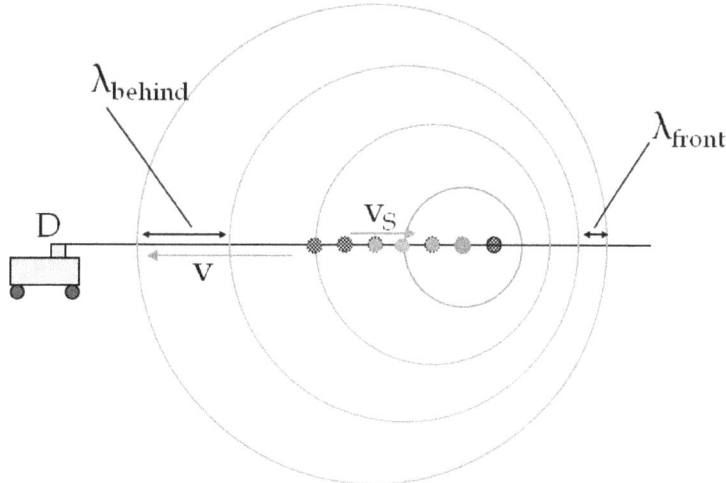

Fig. 4.27: Doppler effect when the detector is stationary while the source is moving. The source sends out wavefronts spaced equally in time at the rate of $1/f_0$ per wavefront. But, these waves are centered at different places because they are generated at the instantaneous position of the source which is changing with time. The distance between the wavefronts in the front would be less than the distance in the back of the wavefront. The wavefronts travel at the same speed in the medium regardless of the state of motion of the source. This means that the detector behind the source will register longer time periods or lower frequencies and a detector in the front of the source will show higher frequency with respect to the frequency with which the waves were emitted at the source.

due to the movement of the source. Hence, to an observer stationary with the medium, the wave fronts will be spaced differently on the two sides of the source. In front of the source, they will appear more closely spaced and in the back more widely separated, because as the source moves, it reduces the distance between a previously emitted wavefront and itself before laying the next wavefront's source point. Therefore, wavelength will be different in front of the source than behind it.

$$\lambda = \begin{cases} (v - v_S)/f_0 & \text{in front of the source, } v_S < v \\ (v + v_S)/f_0 & \text{behind the source} \end{cases} \qquad (4.67)$$

Since the speed of the wave is v in the medium, the frequency observed at the detector will be v/λ .

$$f = \left(\frac{v}{v \pm v_S}\right) f_0 \begin{cases} + & \text{behind the source} \\ - & \text{in front of the source, } v_S < v \end{cases} \qquad (4.68)$$

Both Source and Detector Moving

It is quite common that both source and detector move with respect to the medium. In that case we can imagine a stationary frame of reference in the medium, and then perform a two-step process. First we find the frequency in the imagined stationary frame using the source-moving transformation, and then find the frequency detected by the detector-moving transformation. Let f_0 be the frequency of the source, and v_S, v_D and v be speeds of the source, the detector and wave with respect to the stationary medium. Let f_M be the frequency in the imagined frame stationary with respect to the medium, and f be the frequency observed in the detector.

The following two-step process will give the frequency observed by the detector in terms of the frequency of produced at the source.

$$f_M = \left(\frac{v}{v \pm v_S}\right) f_0 \begin{cases} + & \text{behind the source} \\ - & \text{in front of the source, } v_S < v \end{cases} \tag{4.69}$$

$$f = \left(\frac{v \pm v_D}{v}\right) f_M \begin{cases} + & \text{when detector moves towards the source} \\ - & \text{when detector moves away from the source} \end{cases} \tag{4.70}$$

We can combine these equations to obtain one equation giving us the frequency detected to the frequency produce, as long as we remember the meaning of different signs.

$$f = \left(\frac{v \pm v_D}{v \pm v_S}\right) f_0 \tag{4.71}$$

Note: Because of the nonrelativistic calculation, we cannot use the formulas derived here for light; you will need to use special theory of relativity for correct relation called the **relativistic Doppler effect**. We quote the relation without derivation.

$$f = \sqrt{\frac{c - v_S}{c + v_S}} f_0 \begin{cases} v_S > 0 & \text{if source moving towards receiver} \\ v_S < 0 & \text{if source moving away from receiver} \end{cases} \tag{4.72}$$

4.13 EXERCISES

Coupled Vibrations

Ex 4.1. Two blocks of mass m each are connected by a spring of spring constant k, and then the blocks are placed on a frictionless track, where then can move freely on a straight track. Mark one point on the track as the origin and let the x-axis of a coordinate system be along the track. Let x_1 and x_2 be the positions of the two blocks are arbitrary time. (a) Write the equations of motion of the two blocks. (b) Deduce the equations of motion for the center of mass and the relative coordinate. (c) Is the motion of the center of mass oscillatory? (d) Is the motion of the relative coordinate oscillatory? (e) Find the frequencies of the normal mode(s). (f) What are the expressions of x_1 and x_2 when the relative coordinate oscillates with a frequency ω? (g) What is the energy of the system of the two masses at an arbitrary time?

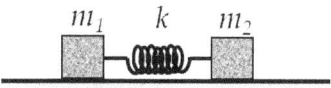

Fig. 4.28: Exercise 4.1.

Ex 4.2. Two blocks of masses m_1 and m_2 are connected by a spring of spring constant k, and then the blocks are placed on a frictionless track, where then can move freely on a straight track. Mark one point on the track as the origin and let the x-axis of a coordinate system be along the track. Let x_1 and x_2 be the positions of the two blocks are arbitrary time. (a) Write the equations of motion of the two blocks. (b) Deduce the equations of motion for the center of mass and the relative coordinate. (c) Is the motion of the center of mass oscillatory? (d) Is the motion of the relative coordinate oscillatory? (e) Find the frequencies of the normal mode(s).

Ex 4.3. Two balls of equal mass m are connected by a spring of spring constant k. The two-ball system is then hung from the ceiling by using another identical spring of spring constant k as shown in the figure. Let the point of suspension be the origin O and the y-axis pointed vertically down. Let y_1 and y_2 be the coordinates of the two balls at some time t. Let l be the stretched lengths of the two springs. (a) Write the equations of motion of the two balls. (b) Let $\bar{y}_1$ and $\bar{y}_2$ denote the y-coordinates of the balls when they are in equilibrium. Find $\bar{y}_1$ and $\bar{y}_2$. (c) Introduce new displacement variables for the two masses by $\eta_1(t) = y_1(t) - \bar{y}_1$ and $\eta_2(t) = y_2(t) - \bar{y}_2$. What are the equations of motions of $\eta_1(t)$ and $\eta_2(t)$? (d) Solving the equations of motion of $\eta_1(t)$ and $\eta_2(t)$ together will give you two normal mode frequencies. Show that the normal mode frequencies are given by $\omega^2 = \left(\frac{3\pm\sqrt{5}}{2}\right)\frac{k}{m}$.

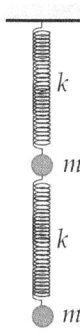

Fig. 4.29: Exercise 4.3.

Wave Function

Ex 4.4. A traveling wave along the x-axis is given by the following wave function

$$\psi(x,t) = 5\cos(2x - 10t + 0.4),$$

where x in meter, t in seconds, and ψ in meters. Read off the appropriate quantities for this wave function and find the following characteristics of this plane wave: (a) the amplitude, (b) the frequency, (c) the wavelength, (d) the wave Speed, and (e) the phase constant.

Ex 4.5. (a) Plot the wave function given in Exercise 4.4 as a function of x at two different times, $t = 0$ and $t = 0.3$ second. These plots are called **wave profile**. (b) Which way is this wave traveling in time?

Ex 4.6. (a) Plot the following two wave functions at $t = 0$, $\psi_1(x,t) = 10\cos(2x - 10t + 0.4)$ and $\psi_2(x,t) = 10\cos(2x - 10t + 0.4 + \pi/2)$, and (b) give an interpretation of the difference in the phase constants of the two waves.

Ex 4.7. (a) Plot the following two wave functions as a function of time at $x = 0$, $\psi_1(x,t) = 10\cos(2x - 10t + 0.4)$ and $\psi_2(x,t) = 10\cos(2x - 10t + 0.4 + \pi/2)$, and (b) give an interpretation of phase constant. In the case of mechanical waves, these plots tell us the vibration of the particles at the origin.

Sound wave

Ex 4.8. The sound produced in the air by a speaker is given by the following pressure differential wave function.

$$\Delta p = (10\ Pa)\cos(kx + 6000t + \pi)$$

where t is in seconds, x in meters and pressure in Pascals. Assume speed of sound in air to be 343 m/s., and density of air to be 1.2 kg/m^3. Find the following quantities.

(a) Angular frequency, (b) Frequency, (c) Wavelength, (d) Amplitude of the wave, (e) The direction the wave is traveling, (f) Bulk modulus of air, (g) Pressure differential at $x = 0$ at $t = 0$, (h) Pressure at $x = 0$ at $t = 0$, (i) Pressure differential at $x = 0$, $t = 1/5000$ sec, (j) Plot the pressure differential at $x = 0$ versus t, (k) Plot the pressure differential at $t = 0$ versus x.

Ex 4.9. The sound produced in air by a speaker is given by the following displacement wave function.

$$\psi(x,t) = (100 \ \mu m)\cos(120x + \omega t + \pi/2),$$

where t is in seconds, x in meters and ψ in micrometers. Assume the speed of sound in air to be 343 m/s and the density of air to be 1.2 kg/m^3. Find the following.

(a) Wavelength, (b) Frequency, (c) Angular frequency, (d) Amplitude of the displacement wave, (e) The direction the wave is traveling in, (f) Displacement of particles at $x = 0$ at $t = 0$, (g) Displacement of particles at $t = 0$, $x = 5$ mm, (h) Displacement of particles at $x = 0$ vs t, (i) Plot pressure differential at $t = 0$ versus x, (j) The pressure differential wave function.

Energy in three-dimensional waves

Ex 4.10. A plane wave in the air of density 1.1 kg/m^3 is given by the following function.

$$\psi(x,y,z,t) = 4\cos\left(10z - 200t + \frac{\pi}{2}\right)$$

where z is in cm, t in sec and ψ in N/m^2.

(a) Find the amplitude of the wave at $t = 0.1$ sec at the points whose coordinates (x, y, z) are given as follows. (i) (0,0,0), (ii) (0,1,0), (iii) (1,0,0), (iv) (1,1,0), (v) (1,1,1), (vi) (-1,1,1), (vii)(1,-1,1), (viii) (-1,-1,1).

(b) What are the wavelength, frequency, and speed of the wave?

(c) What are the intensities of the wave at the following points (i) (0,0,0) and (ii) (1,1,1)?

(d) How much energies will pass through an area 3 cm^2 in 2 sec if the area is in the (i) xy-plane, (ii) yz-plane and (iii) xz-plane?

Ex 4.11. A spherical wave is given by the following function.

$$\psi(r, \theta, \phi, t) = \frac{4}{r}\cos\left(10r - 200t + \frac{\pi}{2}\right)$$

where r is in cm, t in seconds and ψ in N/m^2. Here r, θ, and ϕ are spherical coordinates of a space point (x, y, z) with $r = \sqrt{x^2 + y^2 + z^2}$.

(a) Find the amplitude at following points (x, y, z) at $t = 0.1$ sec. (i) (0,0,0), (ii) (0,1,0), (iii) (1,0,0), (iv) (1,1,0), (v) (1,1,1), (vi) (-1,1,1), (v)(1,-1,1), (vi) (-1,-1,1).

(b) What are the wavelength, frequency, and speed of the wave?

(c) What are the intensities of wave at (i) (0,0,0), (ii) (0,0,1), (iii) (0,0,2), and (iv) (0,0,3)?

(d) How much energies will pass through an area 3 cm^2 in 2 seconds if the area is in spherical shells at following distances from the origin (i) 2 cm, (ii) 4 cm and (iii) 8 cm?

Ex 4.12. A plane wave of wavelength 3 cm and speed 350 m/s travels through a medium of density 1.2 kg/m^3 along the y-axis of a Cartesian coordinate system with an intensity of 49 W/m^2. Write the wave function for this wave.

Ex 4.13. A spherical wave of wavelength 3 cm and speed 350 m/s travels through a medium of density 1.2 kg/m^3. It is found to have an intensity of 49 W/m^2 at $r = 2$ cm from the source. Write the wave function for this wave.

Ex 4.14. A plane wave of wavelength 0.1 mm and speed 6500 m/s travels through a medium of density 8000 kg/m^3 along the x-axis of a Cartesian coordinate system with an intensity of 4900 W/m^2. Write the wave function for this wave.

Decibels

Ex 4.15. The intensity from 10 m from a source of spherical sound waves is measured to be 20 W/m^2. (a) Find the intensity level in decibels. (b) What is the total power emitted by the source?

Ex 4.16. A microphone detects that the sound level in the room is 3 dB with respect to the standard level of sound $I_0 = 10^{-12} W/m^2$. What is the intensity of sound in W/m^2?

Ex 4.17. The intensity of sound drops by half. What is the change in dB?

Ex 4.18. The intensity of sound drops to 10% of its original intensity. What is the change in dB?

Ex 4.19. A person hears a sound level of 40 dB from an air plane engine. (a) If the eardrum has the area of cross-section area of 0.4

cm^2, how much energy does the eardrum receive is one minute? (b) If the sound intensity level goes up to 80 dB, by what factor has the pressure on the eardrum gone up?

Wave properties

Ex 4.20. Find the speed of a mechanical wave in a rope of tension 100 N and mass density 20 g/cm.

Ex 4.21. Find the speed of a mechanical wave in air of density 1.2 kg/m^3 and bulk modulus 10^5 N/m^2.

Ex 4.22. Find the speed of a mechanical wave in water of density 1 g/cc and bulk modulus 2×10^9 N/m^2.

Ex 4.23. A sonar is an ultrasound device that is used to map the surface of the ocean. At a particular place the echo is heard 2 seconds after the sonar sends an ultrasound. How deep is the ocean there? Use 1500 m/s for the speed of sound in salt water.

Superposition of waves

Beats

Ex 4.24. The beat phenomenon is very sensitive for detecting the deviation of sound from a pitch. This helps piano tuners as well as members of orchestra to adjust their musical instruments. When an instrument key and a standard reference for that key do not produce any beats, then the two are said to be in tune. Suppose that a piano key is supposed to be 256 Hz, but it is out of tune. When a tuning fork of 256 Hz is sounded at the same time as the key, one hears a beat of 30 Hz. By how much is the piano key out of tune? Can this information tell you what the frequency of the key is?

Ex 4.25. Two police cars are producing sirens primarily at 2000 Hz when they are not moving. You hear a beat when one of the cars is standing still while the other car is moving towards you at a speed of 40 m/s. Determine the beat frequency. Use the speed of sound in air to be 343 m/s.

Ex 4.26. A piano key of 512 Hz is producing 500 Hz sound. (a) What will be the beat frequency if it is checked against a violin producing

520 Hz? (b) If the wire corresponding to the key is 1.2 meters long, and has a mass density of 0.1 kg/m, by what percentage should the tension of the wire be adjusted so that the pitch is adjusted to 512 Hz? Assume insignificant change in mass per unit length while tightening.

Interference

Ex 4.27. Two coherent microwave sources of wavelength 3 cm are separated by 6 cm. Their electromagnetic waves interfere at a detector 200 cm away. The detector can be moved about in a line parallel to the line joining the two sources. The arrangement is similar to Young's double-slit experiment. Find the locations of the detector the constructive interference for order 0, ±1 and ±2.

Ex 4.28. A light of wavelength 632.8 nm wave is passed through a lens. The beam spreads out and is incident perpendicularly on an opaque board that has two very narrow slits which are separated by 3 μm. The emerging waves from the two slits will interfere constructively and destructively. (a) Find the directions in which the destructive interferences of lowest three orders would take place. (b) At what angle the condition for the constructive interference, based on the small angle approximation, is violated.

Ex 4.29. Two coherent 3-mm wavelength sources separated by 5-mm shine towards a screen as shown. If the intensity of individual source at the screen is 3 W/m^2, what is the intensity at the points P and Q marked on the screen?

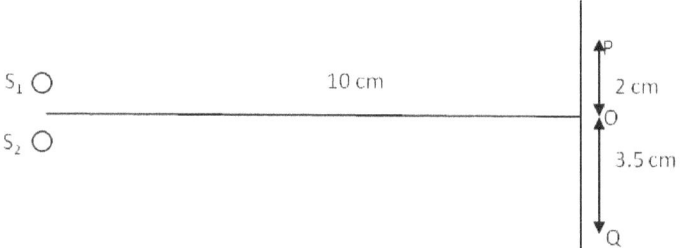

Fig. 4.30: Exercise 4.29.

Ex 4.30. In Exercise 4.29, compute the intensities at points P and Q if the sources are replaced by those of wavelength 2-mm.

Ex 4.31. In Exercise 4.29, let the sources of wavelength 700 nm be separated by 3 micrometer. Find the intensities at the same P and Q as in Exercise 4.29.

Standing Wave

Ex 4.32. A guitar string of length 78 cm and mass per unit length of 30 grams per meter is under a tension of 200 N. (a) Find the frequency of the fundamental mode of vibration. (b) Find the frequencies of the first, second and third overtones. (c) Find the distance between the nodes of the second overtone. (d) Find the distance between nodes of the third overtone.

Ex 4.33. A 2-m long string is under a tension of 300 N. It is found that the frequency of its fundamental tone of vibration is 250 Hz. Find the mass per unit length of the string.

Ex 4.34. A 82 cm string of mass density 20 grams per m is to be tightened to produce a fundamental note of 510 Hz. What must be the tension in the string?

Ex 4.35. A string of length 120 cm, density 20 g/m, and tension 20 N, is vibrating in its first harmonic with maximum displacement of 1/2 cm at any point of the string. (a) Where is/are the antinode(s), and what is the value of amplitude at an antinode? (b) What is the value of the amplitude of oscillations at the following points from one end of the string: (i) 10 cm, (ii) 20 cm, (iii) 40 cm, and (iv) 60 cm? (c) How much time do the particles in (b) take to go from maximum amplitude to zero amplitude?

Doppler effect

Ex 4.36. A car horn at rest sounds a frequency of 400 Hz. (a) You are running towards the horn at a speed of 10 m/s, what frequency will you hear? (b) What frequency will you hear if you were running away from the horn? Assume the speed of sound in air to be approximately 343 m/s.

Ex 4.37. A car horn at rest sounds a frequency of 400 Hz. (a) If the car is moving towards you at speed 10 m/s, what frequency will you hear? (b) If the car is moving away from you at speed 10 m/s, what frequency will you hear? Assume the speed of sound in air to be approximately 343 m/s.

Ex 4.38. While sitting on your porch you hear a fire engine siren at a frequency 1620 Hz when approaching towards you, and a frequency

1590 Hz when receding from you. From this data, determine the speed of the fire engine assuming it was constant. Use 343 m/s for the speed of sound.

Ex 4.39. A humming bird is flying towards you with speed 10 m/s while you are running towards it at 20 m/s, both speeds with respect to the ground. You hear sound of frequency 450 Hz. What is the frequency of the sound the Humming bird is making? Use 343 m/s for speed of sound.

Ex 4.40. A police car emits siren at a frequency of 1000 Hz when at rest. On a highway, you hear the same police car approaching towards you from behind with its siren blowing. You hear 1050 Hz when the car was approaching, and 930 Hz when it has gone past you. Assuming constant speeds of the police car and your car, find their values.

4.14 PROBLEMS

P 4.1. A bat emits an ultrasound of frequency 50,000 Hz which bounces off a wall and returns to the bat in 1 ms. (a) Ignoring the speed of the bat, how far away is the bat from the wall? (b) What is the wavelength of the ultrasound wave. Use density of air 1.3 kg/m^3 and bulk modulus 1×10^5 Pa.

P 4.2. A continuous traveling wave of amplitude 2 cm and frequency 10 Hz rides towards the positive x-axis on a string of mass per unit length 0.1 kg/m and the tension 100 N. At $t = 0$, the displacement at $x = 0$ is found to be 1 cm. Represent the wave by a wave function.

P 4.3. A traveling wave on a string of mass 0.15 kg/m is given by the following wave function.

$$\psi(x,t) = (3 \text{ cm}) \cos\left(0.5x + 1200t + \frac{\pi}{6}\right)$$

where x is in m and t in sec. (a) What is the speed of the wave, and in which direction the wave is traveling? (b) How much is the tension in the string? (c) What is the amplitude of the wave? (d) What is the amplitude of the wave at the origin at $t = 1$ ms?

P 4.4. Two stones are dropped in a pond 3 m apart at the same time. They produce waves of wavelength $\frac{1}{2}$ m. The two waves meet at the rectangular edge of the pond 20 m away, and interfere constructively and destructively. At the symmetric middle point the interference is constructive. Where are the next three constructive interference sites, P_1, P_2, and P_3 as shown in 4.31?

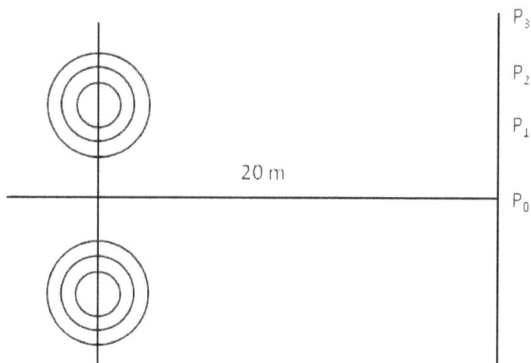

Fig. 4.31: Problem 4.4.

P 4.5. A guitar string of length 60 cm and mass per unit length 10 g/m is tuned to play the note A of frequency 440 Hz at the fundamental. (a) What is the wavelength of the wave on the string? (b) What is the tension in the string? (c) Assuming sound in air travels at speed 344 m/s, what is the wavelength of the sound produced?

P 4.6. A string of length 108 cm is held fixed at one end and the other end is attached to a motor that moves it up and down by 0.3 cm at a set frequency. The tension in the string has a value of 90 N, and it has a mass density of 150 grams per meter. What must be the frequencies of the motor to excite the fundamental mode and the next three harmonics?

P 4.7. Light wave acts similar to the mechanical waves when it comes to interference. Consider a light source that shines a screen far away. When a mirror is brought, as shown in the Fig. 4.32, then a direct ray to the screen interferes with the reflected ray. Note the phase of the wave changes by π radians upon reflection as a result if the difference in distances traveled is an integral multiple of wavelength, the two waves will be completely out of phase and a destructive interference will result there. Similarly, a constructive interference will result where the difference in the distances traveled is odd multiple of $\frac{1}{2}$ wavelength. The interference of the two rays produces bright and dark spots on the screen. The screen is positioned so that the two rays are in phase at the straight point P_0. Find the location of two dark spots (Q_1 and Q_2) and two other bright spots (P_1 and P_2) on the screen.

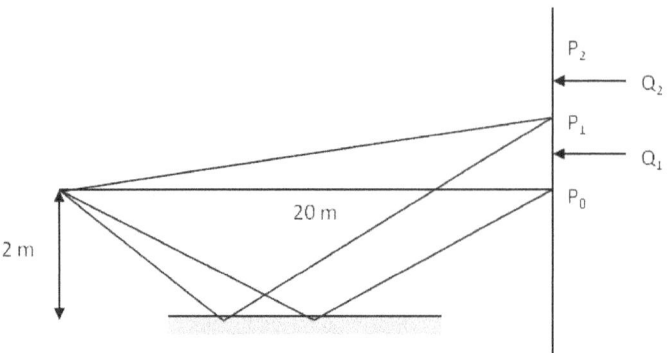

Fig. 4.32: Problem 4.7

P 4.8. A rectangular pulse is given by the following function at time $t = 0$.

$$\psi(x,0) = \begin{cases} 0 & x \leq -1 \\ A & -1 < x < 1 \\ 0 & x \geq 1 \end{cases}$$

The wave is traveling towards the positive x-axis with speed v. (a) What is the wave function at some later time t? (b) Determine the instantaneous power in the wave.

P 4.9. A wave pulse is given by the following function at time $t = 0$.

$$\psi(x,0) = A \exp(-x).$$

The wave travels with speed v towards the negative x-axis. What is the wave function at some later time t?

P 4.10. Two strings on two cellos are tuned to the same frequency of 247 Hz. One of the strings is then tightened further increasing its tension by 30%. When the two cellos are played a beat is heard. Find the beat frequency.

P 4.11. In a valley an echo is heard after 1.5 second of the scream. How far away are the mountains? Use 343 m/s for the speed of sound.

P 4.12. A piano key is to be tuned by using a 440 Hz tuning fork. The key is loose and produces a sound that is lower that 440 Hz. The piano tuner hears 4 beats per second when the key is struck at the same time as the tuning fork. How different is the tension in the cord now compared to what it should be to produce the right frequency?

P 4.13. Two loudspeakers are connected to a 2000 Hz signal generator. A person while standing between the two speakers does not hear any beats, but when he runs from one speaker to another he hears a 25 Hz beat. How fast must he be moving away from one speaker and towards the other speaker? Use 343 m/s for the speed of sound.

P 4.14. An aluminum rod can be vibrated to produce squealing sound in a demonstration called the "singing rod". You can buy an aluminum rod of length approximately 1 m and diameter 6 mm or 3 mm from a hardware store. Balance the rod at the mid-point, and while holding it there, stroke the rod between your thumb and the index finger back and forth from the middle to one end. If you do it right, you will produce a high pitch squeal. Try it. Consider one such rod of length 1.2 m and the area of cross-section 3 mm. (a) What would be the lowest frequency of sound produced? (b) What would be the wavelength of the sound on the aluminum rod corresponding to the lowest frequency sound produced? (c) What is the frequency of the same sound in the air? (d) What is the wavelength of the same sound in the air? Assume the speed of the sound in air to be 343 m/s.

P 4.15. Two speakers connected to the same signal generator are placed 2 m apart. When the signal generator sends sinusoidally varying signal of frequency 500 Hz, the speakers produce pure tone of 500 Hz. The waves coming out of the speakers are in sync with each other. A microphone, which can move parallel to the line joining the speakers, is placed 6 meters from the speakers as shown. An intensity maximum is found when the microphone is in the middle. As the microphone is moved, we find intensity minima and maxima alternately. Find two other intensity maxima (P_1 and P_2) and three intensity minima.

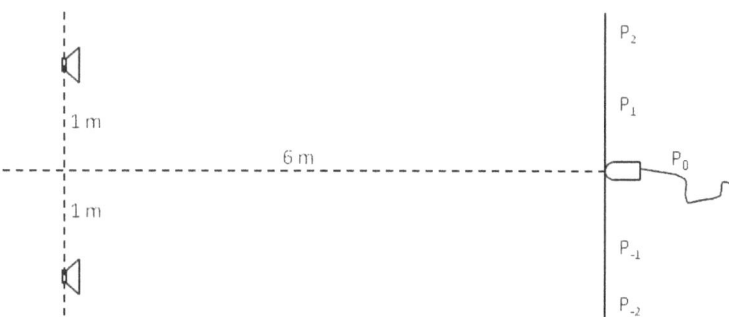

Fig. 4.33: Problem 4.15.

P 4.16. Doppler flow meters work with the principle of the Doppler shift. In these instruments, when sound wave is incident on a moving particle, the reflected wave has an altered frequency because the reflected wave is emitted at the original frequency but the source now is moving with respect to the detector. By detecting the change in the frequency observed by the detector, one can deduce the speed of the particle in the fluid. In a certain experiment, a wave is sent by a transducer, which is a source of ultrasound wave, at 45 degrees angle

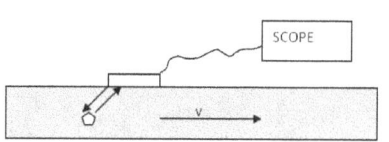

Fig. 4.34: Problem 4.16.

to the flow. The reflected waves are detected by the same transducer that emitted the original wave. When a frequency of 10 MHz is employed, a shift of 1000 Hz is observed. What is the speed of flow? Use the speed of sound in water to be 1500 m/s. [Note the speed of sound in materials is much greater than the speed of sound in the air.]

Chapter 5

STRESS AND STRAIN

Contents

Have you wondered how trusses can support large weights above them? How much weight can be placed before a beam will give way? How far above a ladder is it safe to climb? These and many other questions like them address the balance of forces on and within physical structures. They are the subject of a branch of mechanics called statics. We have already discussed this topic when we introduced forces in an earlier chapter. We will review the salient aspects of statics in this chapter and apply them to understand the internal forces in a body. A study of the static equilibrium helps us with the design of mechanical structures that would be stable for a wide range of conditions, and avoid disastrous situations such as the collapse of a building.

5.1 STATIC EQUILIBRIUM

We say that a system is in a static equilibrium if the net force on every particle of the system is zero. Let there be N particles in a system, and let $\vec{F}_i$ be the net force on the ith particle. Then the **static equilibrium** corresponds to the following set of conditions.

$$\left.\begin{array}{l} \vec{F}_1 = 0 \ \ \text{net force on particle \#1} \\ \vec{F}_2 = 0 \ \ \text{net force on particle \#2} \\ \quad\vdots \\ \vec{F}_N = 0 \ \ \text{net force on particle \#N} \end{array}\right\} \tag{5.1}$$

Recall that the forces on each particle can be classified based on which agent applies that force: if the force is from another particle in the system, we call that force an **internal force** and if the force is by some object outside the system of N particles, then we call that force an **external force**. Suppose we sum up all the internal forces on part 1 and call it $\vec{F}_1^{int}$ and sum up all the external forces on particle 1, call it $\vec{F}_1^{ext}$, then we can write the force $\vec{F}_1$ on particle 1 as

$$\vec{F}_1 = \vec{F}_1^{int} + \vec{F}_1^{ext}.$$

Similarly for the forces on the other particles. Using this separation, Eq. 5.1 should actually be written as

$$\left.\begin{array}{l} \vec{F}_1^{int} + \vec{F}_1^{ext} = 0 \ \ \text{net force on particle \#1} \\ \vec{F}_2^{int} + \vec{F}_2^{ext} = 0 \ \ \text{net force on particle \#2} \\ \quad\vdots \\ \vec{F}_N^{int} + \vec{F}_N^{ext} = 0 \ \ \text{net force on particle \#N} \end{array}\right\} \tag{5.2}$$

Ideal nondeformable objects and static equilibrium

All physical bodies are deformable and are deformed when a force is applied on them. Even when an object does not appear deformed, there must be compression or stretching at the microscopic level, otherwise we would not be able to explain the development of reaction forces such as the normal force and the frictional force. For instance, when you place a book on the table, neither the book nor the table appears deformed, but they must be a little compressed at the contact surface so that a normal force can develop there.

For the sake of mathematical simplicity, we will first study "ideal" nondeformable objects defined as follows. An ideal nondeformable object applies reaction forces such as the normal and friction forces and action-at-a-distance forces on the external bodies but its shape does not change under the prevailing conditions. Most solids and liquids will fall into this category.

Suppose you have an ideally strong nondeformable object. Then, the requirement of vanishing of forces on all parts can be shown to simplify to a consideration of only the external forces. When you sum up the equations for force on each particle given in Eq.5.2, the internal forces will cancel out since they are all paired up as demanded by the Newton's third law of motion.

$$\vec{F_1}^{ext} + \vec{F_2}^{ext} + \cdots + \vec{F_N}^{ext} = 0 \tag{5.3}$$

Now, if you pick an arbitrary pivot point, and calculate the net torque about that point, you will find that the net torque is also zero if the internal forces between two particles act along the line joining them.

$$\vec{\tau_1}^{ext} + \vec{\tau_2}^{ext} + \cdots + \vec{\tau_N}^{ext} = 0 \tag{5.4}$$

Therefore, for a nondeformable object, the vanishing of net external force and net external torque assures static equilibrium.

For simplicity, all the problems considered in this chapter will have forces in one plane only, which we will take to be the xy-plane. Therefore these two conditions for the static equilibrium would yield at most three equations for an analysis: two for the x and y-components of the net external force and one for the net torque about an axis parallel to the z-axis passing through an arbitrarily chosen pivot point.

$$F_{1x}^{ext} + \vec{F}_{2x}^{ext} + \cdots + \vec{F}_{Nx}^{ext} = 0 \tag{5.5}$$

$$F_{1y}^{ext} + \vec{F}_{2y}^{ext} + \cdots + \vec{F}_{Ny}^{ext} = 0 \tag{5.6}$$

$$\tau_{1z}^{ext} + \tau_{2z}^{ext} + \cdots + \tau_{Nz}^{ext} = 0 \tag{5.7}$$

Deformable objects and static equilibrium

For an arbitrary structure to be in a static equilibrium, the vanishing of the net external force does not ensure static equilibrium. For instance, if you push on a piece of foam from two opposite sides, the net force on the foam is zero, but the foam gets deformed because of an imbalance between the external and internal forces on the particles at the surface.

While there is no limit on external forces, the internal forces, just like the static frictional force, have an upper limit and can be overcome by a large enough external force. Therefore, to study the equilibrium of a deformable object, we need the complete condition for the static equilibrium given in Eq. 5.2, viz., the net force on each and every particle of the system must vanish independently.

To be in static equilibrium, the external forces must balance the internal forces at each point. In this way, the analysis of the static equilibrium of deformable systems provides us with an understanding of the strength of the material. In this chapter, we will study equilibrium with regard to the external forces as well as internal forces.

But, for simplicity, we will look at the problems for the ideal undeformable objects first. This would be a review of our treatment of the static equilibrium in the chapter on forces.

5.2 SOLVING STATICS PROBLEMS

Let us consider simple examples that help us understand the use of the static equilibrium in "ideal" undeformable objects. No object in reality is an undeformable object. But, if the applied forces are not large enough to overwhelm the internal forces, then, the "ideal" system analysis gives us most of the physics and is an important first step in any stability analysis. The method for solving static problems follow a few basic steps outlined below.

1. Pick the system. It can be either the entire structure, a part of the structure, or some small mass element inside the structure. The choice depends on the physical question under study.

2. Identify forces on the system by the external agents, making particular note of where they act and what their directions are. These usually fall into two categories:

   (a) Action-at-a-distance forces, usually gravity in our case, which can be taken to act at the center of mass of the system, and

(b) Contact forces, usually a combination of normal and static frictional force but there may be other forces as well.

3. Draw a free-body diagram and choose x and y-axes judiciously for generating the F_x and F_y equations.

4. For the torque equation, you will need to go back to the original figure where you have both the location and the direction of forces. Pick a reference point O, so that torques are easy to calculate. A good location for O usually makes some of the torques zero for forces passing through O, especially for forces that are unknown. Use lever-arm method rather than the cross-product to calculate torques, and employ clock-wise and counter-clockwise sense of torques to assign plus and minus signs to the individual components of the torques. The algebraic sum of all torques is then set to zero.

5. The three equations given in Eq. 5.5-5.7 are then solved together. you may not need all three equations for every problem, so you would need to understand what you need to do for the particular problem at hand.

We will repeat the examples of this method presented in the chapter on forces. If you have already studied them, then this will be just a review. But, if you skipped those parts of the book, then here is your chance to study them before you proceed to more advanced topics in this chapter.

Example 5.1. Safety on a ladder. *A ladder of length L and of mass M slides when the ladder leans against a frictionless wall at an angle greater than θ from the ground. What is the minimum static frictional coefficient between the ground and the ladder?*

Solution. Let us first draw a picture of the situation and locate all the forces on the ladder. Then we will extract a free-body diagram from this diagram for writing out the equations corresponding to the balance of the external forces. To calculate the torque of the forces, the free-body diagram is not useful. For the torque calculation, we will go back to the original force diagram and choose a convenient pivot point. The choice of pivot point is arbitrary since the net force is zero.

The diagram of forces on the ladder shows various external forces on the ladder and the place those forces act on the ladder. This is given in Fig. 5.1. Note that there are two normal forces on the ladder, one from the floor and the other from the wall. Similarly, there are two frictional forces on the ladder, one from the floor and the other from the wall. The problem states that the frictional force from the wall is zero.

One more thing to note here is that I had make a decision about the direction of the static frictional force. Since, the ladder is not accelerating with respect to the ground, I anticipated that this frictional force must balance the normal force from the wall, and that is why I choose the frictional force to be in that direction.

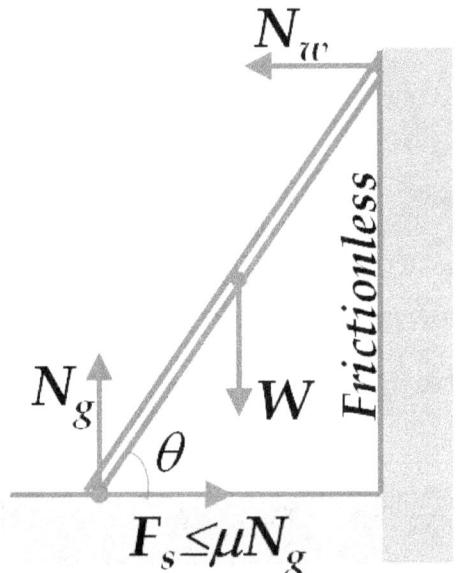

Fig. 5.1: Example 5.1.

In the free-body diagram of forces, we place the forces "coming out" from the same place even though the forces actually act at different points of the body. The free-body diagram of the forces and the direction of the axes are shown in Fig. 5.2 where we also show the directions of the positive x and y-axes. As you may already know that we do not need to pick the position of the origin for a free-body diagram since we use the free-body diagram to write the components of forces with respect to the axes.

Since the acceleration of the ladder is zero, the following equations are immediately found from the free-body diagram. **Note that the relations that include static frictional forces are inequalities, not equalities since the frictional force can take values up to a maximum value.** In the present problem, we will write an equality since the problem states that maximum frictional force should be acting since we are looking for the minimum value of μ.

$$x \text{ equation: } N_w - \mu N_g = 0 \tag{5.8}$$
$$y \text{ equation: } N_g - W = 0 \tag{5.9}$$

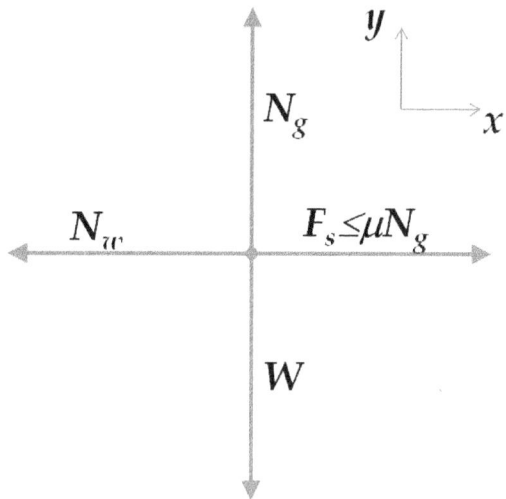

Fig. 5.2: Example 5.1.

Note that we have three unknowns, N_g, N_w and μ, and we have only two relations so-far. The third relation is obtained when we apply the vanishing of torque. The torque of a force depends on the force and the lever arm of that force, which is the distance from the pivot point to the line of action of the force. The information about where the force is acting is lost in the free-body diagram. Therefore, we need to go back to the original diagram given in Fig. 5.1 to examine the situation regarding the torques.

Since the net force is zero, the choice of pivot point would be arbitrary. In the static problems, we choose a pivot point so that the lever arm for some of the unknown forces would be zero for that choice. In the present case, we notice that if we choose the pivot point at the point where the ladder touches the ground, and then the torques from two of the forces, viz, N_g and $F_s = F_s^{\text{max}} = \mu N_g$ will be zero. The digram for the torque calculation is displayed in Fig. 5.3, where the lever arms OC for the force $\vec{W}$ and OD for the force $\vec{N}_w$ are also shown. Let us denote these lever arms by L_w and L_{nw} respectively. The z-axis is coming out of the page in this figure. We find the z-components of the torques and equate their sum to zero for the equilibrium condition to obtain the third relation we seek.

$$N_w\, L_{nw} - W\, L_w = 0. \tag{5.10}$$

The lever arms can be written in terms of the length L of the ladder and the angle θ the ladder makes with the horizontal.

$$L_w = \tfrac{L}{2}\cos\theta \tag{5.11}$$
$$L_{nw} = L\sin\theta \tag{5.12}$$

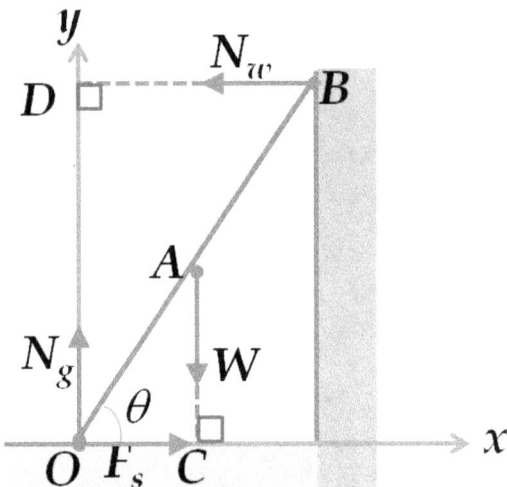

Fig. 5.3: Example 5.1.

Substituting in the torque equation, Eq. 5.10, we find the following.

$$N_w \, \sin\theta - \frac{W}{2}\cos\theta = 0. \qquad (5.13)$$

Now, we can solve Eq. 5.8, 5.9, and 5.13 together to find the result.

$$\mu = \frac{1}{2}\cot\theta. \qquad (5.14)$$

Example 5.2. Cantilever. *A beam that extends beyond its support*

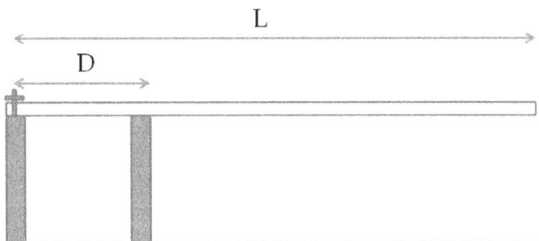

Fig. 5.4: Example 5.2.

is called a **cantilever**. *Consider a cantilever of length L and mass M resting and bolted to two supports, one at the end and the other is at a distance D from it. (a) Find the conditions on the forces such that the cantilever is in static equilibrium. (b) For L=30 m, M=1000 kg, and D=12 m, find the magnitude and directions of the forces on the beam from the supports.*

Solution. We will consider the beam of the cantilever as the system. This makes everything else external to the system. The beam has the forces from the support and the bolt at one end, an upward normal force from the support in the middle, and the force of gravity from the Earth. As shown in Fig. 5.5, we will combine the forces of the bolt and the support at the end into a horizontal force $\vec{F}_{11}$ and a vertical force $\vec{F}_{12}$, the sum of the two forces at the end give us the net force $\vec{F}_1$ at that point.

$$\vec{F}_1 = \vec{F}_{11} + \vec{F}_{12}.$$

Normally, one writes the force $\vec{F}_{11}$ as $\vec{F}_{1x}$ and $\vec{F}_{12}$ as $\vec{F}_{1y}$. This notation is confusing since symbols F_{1x} and F_{1y} stand for x and y-components of the vector $\vec{F}_1$ and are not vectors themselves. Therefore, we will use $\vec{F}_{11}$ and $\vec{F}_{12}$ for the the horizontally and vertically pointed vectors whose sum is the force at the end.

Based on physical grounds we expect that all four forces $\vec{F}_{11}$, $\vec{F}_{12}$, $\vec{F}_2$ and the weight $\vec{W}$ will be in one plane, which we take to be the xy-plane of the coordinate system shown.

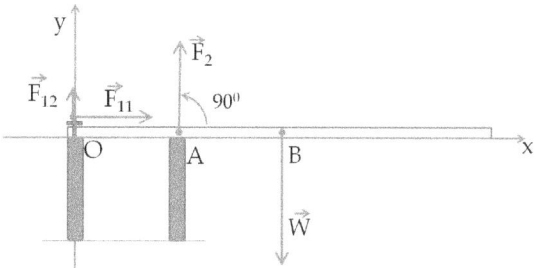

Fig. 5.5: Example 5.2. Forces on the cantilever beam.

Balancing the forces immediately gives the following equations.

$$F_{11} = 0 \qquad\qquad (5.15)$$
$$F_12 + f_2 - W = 0 \qquad\qquad (5.16)$$

Although you can do the torque part graphically, I will use the analytic approach just to illustrate the kind of considerations that goes into this method. For the torque equation, we need to choose a suitable point P about which we will work out the torques. As the net force is zero, the choice of a pivot point is arbitrary. Since we do not know either $\vec{F}_1$ (or equivalently $\vec{F}_{11}$ and $\vec{F}_{12}$) or $\vec{F}_2$, it does not save us much work whether we place the pivot point at O or A. Let us us pick O as the pivot point because it is nicely on one end so that we can measure displacements of the forces from the end point. We will also name the displacements as $\vec{r}_1$, $\vec{r}_2$ and $\vec{r}_w$ for the four forces. Let us organize the information regarding forces and displacements in a table.

Force	Force in components	Displacement	Torque
$\vec{F}_{11}$	$F_{11}\hat{u}_x$	$\vec{r}_1 = 0$	0
$\vec{F}_{12}$	$F_{12}\hat{u}_y$	$\vec{r}_1 = 0$	0
$\vec{F}_2$	$F_2\hat{u}_y$	$\vec{r}_2 = D\hat{u}_x$	$DF_2\hat{u}_z$
$\vec{W}$	$-Mg\hat{u}_y$	$\vec{r}_2 = (L/2)\hat{u}_x$	$-(L/2)Mg\hat{u}_z$

Balancing the torques gives the following equation.

$$DF_2 = \frac{1}{2}LMg. \tag{5.17}$$

Equations 5.15, 5.16, and 5.17 can be solved to yield the following for the magnitudes of the forces.

$$F_{11} = 0 \tag{5.18}$$

$$F_{12} = Mg\left(1 - \frac{L}{2D}\right) \tag{5.19}$$

$$F_2 = \left(\frac{L}{2D}\right)Mg \tag{5.20}$$

Equation 5.18 says that there is no horizontal component of force $\vec{F}_1$ at the end, and Eq. 5.19 says that if the second support is less that $L/2$ from the end, then the force in the vertical direction at the end support is downward, i.e. the force there is dominated by the push back from the bolt at the top of the beam.

(b) Plugging in the numerical values we find the following.

$$F_{11} = 0,$$
$$F_{12} = -2500 \ N,$$
$$F_2 = 12000 \ N,$$

where I have rounded off the magnitudes to the two significant digits.

5.3 INTERNAL FORCES, ELASTICITY AND STRESS

In the last section we studied the equilibrium among external forces. If a system was perfectly rigid and not deformable, the story would end there. But, real materials respond to the external forces, and they are deformable to some extent, even to the point of breaking if the external force is sufficiently large. However, when the deformation is small compared to the physical dimensions of the object, many materials often recover the original shape when the external force is removed. This property is called **elasticity**.

The elastic behavior of a material can be seen by attempting to elongate a wire by hanging masses to the wire as shown in Fig. 5.8.

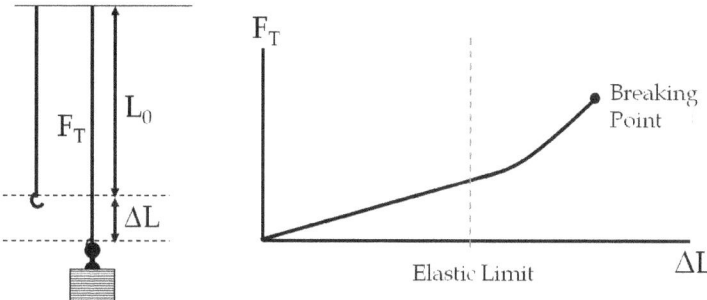

Fig. 5.6: Elasticity of a string. The string is stretched as massed are added. The wire will go back to the origin length when you remove the weights if the stretching is within the elastic limit. If the wore is stretched to breaking point, the wire will break. Between the stretching limit and the breaking point, the wire becomes permanently stretched.

Initially as the weight is increased, the length of the string expands in proportion to the weight. This is the elastic region, since, if the weight is removed, the wire will go back to its original length. If you continue to increase the weight past a certain point, called the **elastic limit**, the elongated wire will not go back to the original length L_0. Instead, the wire will be permanently elongated. Now, if you continue to add more weight to the wire, eventually the wire will break at some point, called the **breaking point**. The breaking point gives us information about the ultimate strength of the material. The region between the elastic limit and the breaking point is called the **plastic region**. We will be concerned with the elastic range here since we want a predictable linear restoring force to help analyze the equilibrium behavior of structures so as to prevent permanent deformation or fracture.

In the linear elastic regime, the magnitude of the force $\vec{F}$ applied to the wire is proportional to the elongation ΔL.

$$F = k\Delta L, \tag{5.21}$$

where k is the proportionality constant. This observation was first made by Robert Hooke (1635-1703), a contemporary of Isaac Newton, and is often called Hooke's law, which states that "Ut tensio sic uis," in Latin, meaning "As the extension, so the force". We have already encountered Hooke's law in the context of our discussions on the spring and tension forces. The reason for a material "to act like" springs can be traced to the inter-atomic bonds in materials as illustrated in Fig. 5.7, which have a spring-like behavior for small displacement from equilibrium.

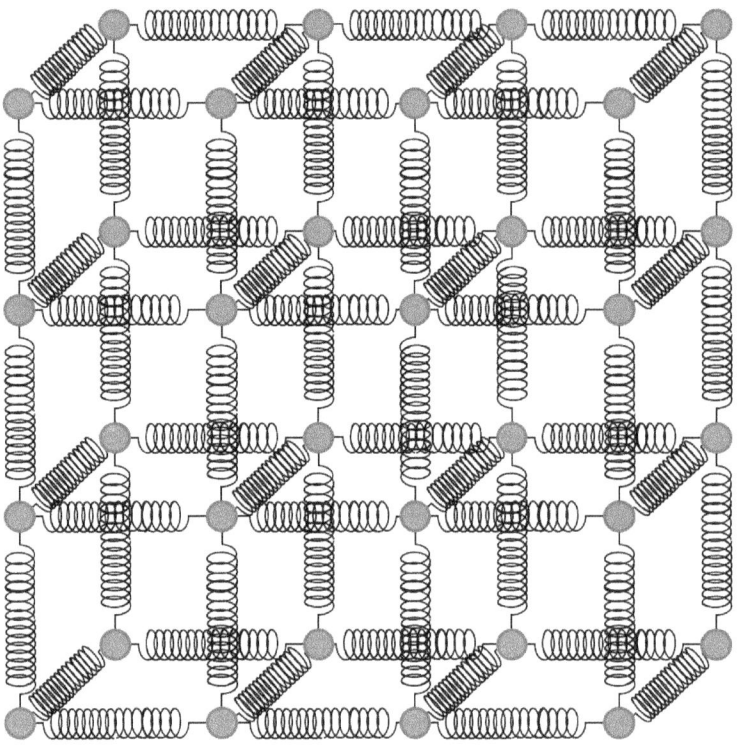

Fig. 5.7: The elastic properties of materials emerge from spring-like bonding of molecules inside the material as illustrated here.

To measure the amount of deformation from the initial length as a fraction, it is customary to introduce a quantity, called the **strain** or **relative deformation**, by the following relation:

$$\text{Strain} = \frac{\text{Deformation}}{\text{Original Length}} = \frac{\Delta L}{L_0} \tag{5.22}$$

Clearly, strain is a dimensionless quantity since it is a ratio of two lengths. Experiments show that, for the same force, a wire stretches less if it is thicker. Therefore, the effectiveness of a force to cause strain in a material is contained in force per unit cross-sectional area of the wire rather than the force alone. To describe the cause of the strain, we define a quantity, called **stress**, which is equal to force per unit area. The magnitude of stress is given by

$$\text{Stress} = \frac{\text{Magnitude of Force}}{\text{Are of cross-section}} = \frac{F}{A}. \tag{5.23}$$

Note that stress will have direction since force has direction. Even the area of cross-section has an orientation. For instance, the six sides of a cube has different orientations in space. Therefore, in general, the "direction" of stress is dependent on the direction of the force and the orientation of the area. The stress is a tensor and has complicated rules for the direction. We will not delve into the general properties of stress here. We will be content to work with the stress given as the force on one area.

The dimension of stress can be found from the ratio of the dimensions of force and area.

$$[\text{stress}] = \frac{[F]}{[A]} = \frac{[M][L]/[T]^2}{[L]^2} = \frac{[M]}{[L][T]^2}. \tag{5.24}$$

This gives the unit of stress to be N/m^2, which is also called Pascal denoted by P. Now, Hooke's law given in Eq. 5.21 can be written using the stress and strain by dividing both sides of the equation by the area A and introducing the factor L_0/L_0 on the right side.

$$\frac{F}{A} = \frac{k}{A}\Delta L = \left(\frac{kL_0}{A}\right)\frac{\Delta L}{L_0} \tag{5.25}$$

Writing the constants given in parenthesis as one letter Y, we have the Hooke's law for a change in length of the wire as

$$\text{Stress} = Y \times \text{Strain} \tag{5.26}$$

The proportionality constant Y is called the Young's modulus. The unit of Young's modulus is the same as that of stress, i.e., N/m^2 or Pascal, since strain in unitless. Table 5.1 lists Young's modulus of a number of common materials.

Example 5.3. Stress inside a hanging wire. *The stress at a point inside the wire is caused by hanging masses as well as the mass of the wire itself. In this example, I will work out the stress at a point P inside a wire of length L and mass m when a mass M is hung from the bottom as shown in Fig. 5.8.*

Let us use a coordinate system with the y-axis pointed up. Let y be the coordinate of the point P from the origin which is at the floor. Let h be the y-coordinate of the lowest part of the wire when the length of the wire is L. Let us examine the balancing of forces on the element of the wire between y and $y + \Delta y$. The forces on this element of wire are tension pointed up above the element, which has the magnitude $T(y+\Delta y)$, the tension below the element pointed down, which has the magnitude $T(y)$, and the weight of the element pointed down, which has the magnitude $m\Delta y/L$. Since the acceleration of the element is zero, the forces are balanced. This gives us the following equation.

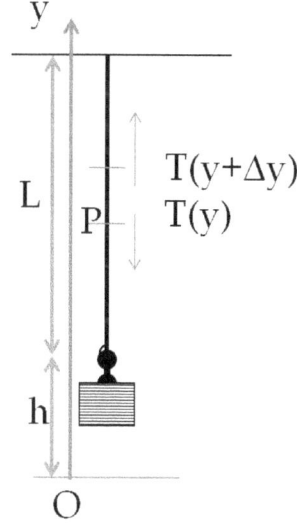

Fig. 5.8: Example 5.3.

Table 5.1: **Young's, shear and bulk moduli**
Source: Kaye and Laby, Tables of physical & chemical constants, 16th edition
(1995). Young's Modulus Y, Shear Modulus G, Bulk Modulus B. (Temp=

Material	Y (GPa)	G (GPa)	B (GPa)
Metals and alloys			
Aluminum	70.3	26.1	75.5
Copper	129.8	48.3	137.8
Lead	16.1	5.59	45.8
Nickel (unmagnetized., hard)	219	83.9	187.6
Silver	82.7	30.3	103.6
Titanium	115	43.8	107.7
Tungsten	411.0	160.6	311.0
Brass (70 Zn, 30 Cu)	100.7	37.3	111.8
Steel, Stainless	215.3	83.9	166.0
Tungsten Carbide	~ 534.4	~ 219	~ 319
Glass and plastics			
Glass, Heavy Flint	80.1	31.5	57.6
Glass, Crown	71.3	29.2	41.2
Quartz, fused	73.1	31.2	36.9
Epoxy	~ 3.2		
Polycarbonate	2.4		
Polyethylene	0.4-1.3		
Polyvinylchloride (PVC)	2.4-4.1		
Liquids Glycerin		20.5	
Mercury		26.2	
Olive oil		1.60	
Paraffin oil		1.62	
Turpentine		1.28	
Water (150)		2.05	
Water, sea		2.32	

(to the left of the Glass and plastics heading: $20°C$))

$$T(y + \Delta y) - T(y) - \frac{m\Delta y}{L}g = 0, \quad h \leq y \leq h + L. \qquad (5.27)$$

Rearranging and dividing by Δy, and taking the infinitesimal Δy limit we obtain the following equation for $T(y)$.

$$\frac{dT}{dy} = \frac{mg}{L}, \quad h \leq y \leq h + L. \qquad (5.28)$$

We can solve this equation for the tension in the wire at point P by rewriting this equation as $dT = (mg/L)dy$ and integrating from $y = h$ to $y = y$.

To specify the values of tension at the two limits in this integration, we need the value of tension when $y = h$, i.e. at the bottom point of the string. To obtain this value, we will apply Newton's second law to the element of the wire that is at that point. This element of length Δy has a tension pulling it up with magnitude T_h (which we wish to determine), weight of the element $mg\Delta y/L$, and the weight of the hanging mass of magnitude Mg pointed down. Therefore,

$$T(h + \Delta y) - \left[M + \frac{m}{L}\Delta y\right]g = 0. \qquad (5.29)$$

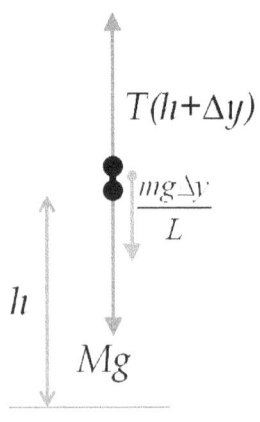

Fig. 5.9: Free-body diagram for an element at the tip of the wire.

We are interested in the infinitesimal limit of Δy so that we can get the information at the tip of the wire. This will leave only M in the bracket and $T(h + \Delta y)$ will go into $T(h)$, which we will write as T_0 to indicate that T_0 is the tension at the bottom point of the wire. This gives

$$T_0 = Mg. \tag{5.30}$$

Now, the integration of Eq. 5.28 goes as

$$\int_{T_0}^{T(y)} dT = \int_{h}^{y} \frac{mg}{L} dy, \quad h \leq y \leq h + L. \tag{5.31}$$

This equation along with Eq. 5.30 gives the result we seek.

$$T(y) = Mg + \frac{mg}{L}(y - h). \tag{5.32}$$

This result shows that if the mass of the wire cannot be neglected compared to the mass hung from the wire, then, the tension in the wire will vary along the wire.

5.3.1 Types of Stress

Tensile versus Compressive Stress

A stress that causes strain in the same direction as the applied force can lead to either the extension of the body or a compression of the body as illustrated in Fig. 5.10. The stress that leads to an extension of the body is called a **tensile stress** and the stress that compresses the body is called a **compressive stress**.

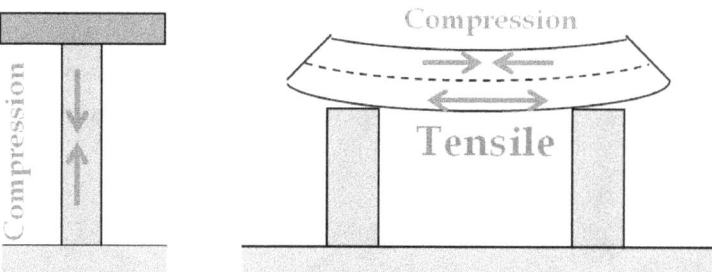

Fig. 5.10: If the strain in the material is negative, i.e. the material is compressed as a result of the stress, then the stress is called a compressive stress. If the strain in the material is positive, i.e. the material stretches as a result of the stress, then the stress is called a tensile stress.

Usually Young's moduli for the two stresses are equal, but they do not have to be equal. For instance, the Young's modulus for concrete is 2 MPa for the tensile stress and 20 MPa for the compressive stress. Thus, a column made of concrete will break more easily when pulled than when compressed; a concrete column is able to support considerably more weight on the column than it can hold up weight hanging from it.

Shear Stress

When a force is applied parallel to the surface of an object, instead of stretching or compressing the object, the force may deform the object if some other part of the object is held fixed. For instance, suppose you fix the bottom of a box by gluing the bottom part and apply a side-ways force on the top part as illustrated Fig. 5.11, you will deform the box. This deformation is used to define an **angular or shear strain**.

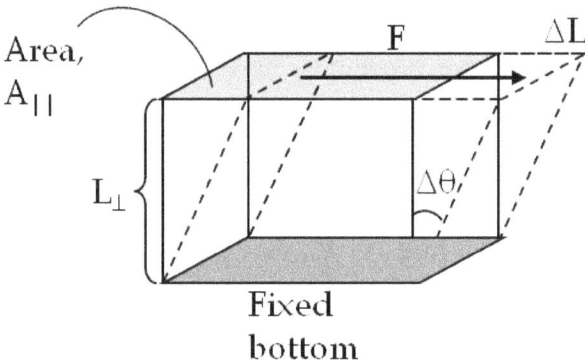

Fig. 5.11: Defining shear strain and shear stress. The bottom part of the box is fixed to a table and a force parallel to the top surface is applied. The top face shifts by a distance ΔL compared to the bottom face. The shear strain is defined by the ratio $\Delta L/L_\perp$, where $L_\perp$ is the distance between the shifted face and the fixed face. Shear stress is defined as F/A, where A is the area of the shifted face.

The shear strain is defined by the ratio $\Delta L/L_\perp$, where ΔL is the displacement of the shifted face compared to the fixed face and $l_\perp$ is the distance between the shifted face and the fixed face. For small deviations, the angle of the deviation $\Delta\theta$ is equal to $\Delta L/L_\perp$, and is a measure of the shear strain.

$$\text{Angular or shear strain} = \frac{\Delta L}{L_\perp} \approx \Delta\theta. \qquad (5.33)$$

The resulting stress in the body is called the **shear stress**. This time, the applied force is spread over the area of the surface parallel to the force and not perpendicular to the force, as was the case for the tensile and compressive stresses. Therefore, the stress corresponding to a shear strain is defined as the magnitude of the force $\vec{F}$ divided by parallel area, $A_\parallel$.

$$\text{Shear stress} = \frac{F}{A_\parallel}. \qquad (5.34)$$

Hooke's law is also applicable for this situation if the angular deviation is small. That is, we expect the shear stress to be proportional to the shear strain.

$$\text{Shear stress} = G \times \text{Shear strain} \qquad (5.35)$$

where the proportionality constant G is called the shear modulus of the material. The shear modulus G has the units of N/m^2 or Pa. The shear moduli of some common materials are listed in Table 5.1. As a general rule, the shear modulus is less than the Young's modulus for the same material, which means that it is usually easier to deform an object sideways than to compress or elongate it.

Note that you must pay attention to the particular area and length in the definition of shear stress given in Eq. 5.34. Unlike the tensile stress, the area here is of the surface that is parallel to the applied force. Further more, when you find the angle of deviation to use in the formula for the the angular deviation, you need to make sure the displacement ΔL and the length $L_\perp$ are perpendicular to each other. Also, while ΔL is parallel to the applied force, $L_\perp$ is perpendicular to the applied force.

Ultimate strength

I have mentioned above that if the magnitude of the applied force exceed a certain value, the material will ultimately break up. This was illustrated in Fig. 5.8 for stretching of a wire. We expect the same physics should happen for the compressive, tensile and shear stresses.

The compressive, tensile and shear stresses lead to three kinds of fractures in materials, called tensile fracture, compressive fracture and shear fracture, as displayed in Fig. 5.12.

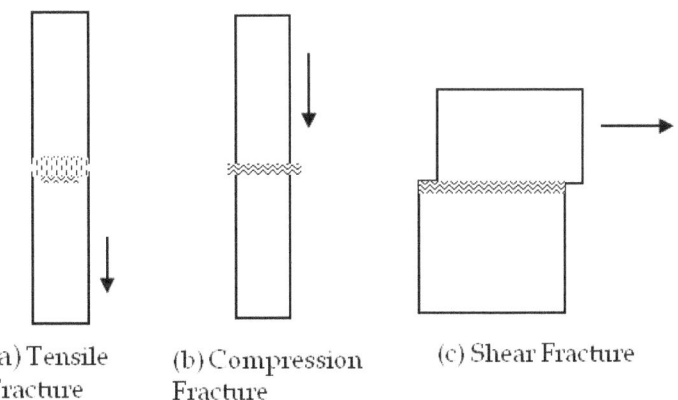

(a) Tensile Fracture   (b) Compression Fracture   (c) Shear Fracture

Fig. 5.12: Compressive, tensile and shear fractures.

Note than the shear stress rupture is in the perpendicular dimension to the direction of the shear force while the tensile or compression related ruptures take place in the same direction as the corresponding force. The ultimate stress of each type that a material can withstand before giving in is called the **ultimate strength** of the material.

Whenever you design a mechanical structure, you must take into account the ultimate strength of the material to be used and the type of stress to be expected under extreme conditions of use. This is especially important when designing the structures that are subject to large stresses such as buildings and bridges. The ultimate strength for the shear stress is usually less than that for compressive and tensile stresses as we see in Table 5.2 and therefore, a building is more susceptible to a shear stress than to compressive or tensile stress.

Table 5.2: Ultimate strength of materials (MPa) Source: Various

Material	Tensile strength (MPa)	Compressive strength (MPa)	Shear strength (MPa)
Aluminum	90-100		
Brass, rolled	230-270	~ 250	~ 200
Concrete	~ 3	20-30	~ 2
Iron, cast	100-230	~ 550	~ 170
Iron, wrought	290-450		
Steel	400-1500	~ 500	~ 250
Wood, pine (parallel to grain)	20-50	~ 35	~ 5

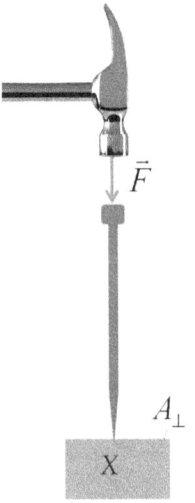

Fig. 5.13: The pressure at the point marked X is the magnitude F of the force divided by the area $A_\perp$ of the tip of the nail.

Pressure

We have looked at stresses caused in a material by forces applied perpendicular to the opposite faces of the material and those applied parallel to the surfaces of the material. Now, we consider forces applied on the system from all directions, as would be the case, for instance, when an object in submerged in a fluid, where fluid will apply the force from all directions.

Pressure at a point is defined as the magnitude of the force per unit area. The area is the area perpendicular to the force and we can indicate this in the formula for pressure. For instance, if you apply a force on a nail, as illustrated in Fig. 5.13, then the area will be the area of cross-section of the tip of the nail. Suppose a force $\vec{F}$ acts on an area $A_\perp$. Then the pressure corresponding to the force will be defined as

$$\text{Pressure, P} = \frac{F}{A_\perp}. \tag{5.36}$$

Formally, this equation looks identical to the equation for the definition of compressive and tensile stresses. However, when we speak of pressure, we are usually interested in a different strain than the change in linear dimension. In the case of pressure as a stress, we are interested in the change in volume. The strain of interest for the pressure is the relative change in the volume, which is also called the **bulk strain**, which is defined as the change in the volume ΔV divided by the original volume.

$$\text{Bulk strain} = -\frac{\Delta V}{V_0}. \tag{5.37}$$

Hooke's law also holds for the pressure and the bulk strain

$$\boxed{\text{Pressure} = B \times \text{Bulk strain,}} \tag{5.38}$$

where the proportionality constant B is called the **bulk modulus**. When you apply a force from all sides of an object, the volume of the object will change and a pressure will develop inside the object.

Example 5.4. Stress in a steel beam. *A steel beam of length L and area of cross-section A is posted vertically on a concrete slab. Find the tensile stress in the beam as a function of height h.*

Solution. We have done a similar problem for a wire under stress. Same technique will give the stress at any point equal to the weight of the steel beam above that point. Suppose the origin of the y-coordinate is at the top point of the beam and the direction of the y-axis is pointed down. Then, the weight of the beam above the point with y-coordinate equal to y will be My/L. Therefore, the stress at that point will be

$$\text{Stress at a distance } y \text{ from the top} = \frac{M}{L}y.$$

We can state the answer for the height from the bottom as.

$$\text{Stress at a height h from the bottom} = \frac{M}{L}(L - h).$$

Example 5.5. Compression of a support beam. *Two cylindrical concrete posts are to support a heavy machine of mass 20,000-kg set in the middle of a 2000-kg beam resting on the posts. For safety reasons, it is required that the stress in the posts does not exceed 5 times the ultimate strength of concrete, what must be the minimum radii of the two posts.*

Solution. The forces on the beam are given in Fig. **??**. In the figure M and m are the masses of the machinery and the beam respectively. The force $\vec{F}_1$ and $\vec{F}_2$ are the forces on the beam from the two posts respectively. By symmetry in the situation, the two forces must be equal. You can prove this also by demanding that the net torque be zero. Let us represent both of them by $\vec{F}$.

$$\vec{F}_1 = \vec{F}_2 \equiv \vec{F}.$$

From the static condition for the beam we immediately obtain the magnitude of the force $\vec{F}$.

$$F = \frac{1}{2}(M + m)g.$$

This gives the compressive force in each column, just from the weight of the machinery and the beam. There is also compressive stress in each support from the weight of the supports above the point of consideration.

Suppose the stress in the support is largely from the machinery and the steel beam. Then, the stress in each support will be F/A, where A is the area of cross-section of the support. Now, if we equate the compressive stress to safety factor c times the ultimate strength for compressive stress.

$$\frac{F}{A} = c \times \text{Ultimate strength}$$

This gives the following for the area of cross-section of the support we need.

$$A = \frac{F}{c \times \text{Ultimate strength}}.$$

We look up the ultimate strength of concrete for compression mode in the table. This says that the ultimate strength $= 20 \times 10^6$ N/m^2. Therefore, to be safe, the area of cross-section that meets the safety requirement is $A = 0.027$ m^2. Therefore minimum radius of the posts must be 9.25 cm^2.

Example 5.6. Tear under shear stress. *A steel bolt is used to connect two steel plates. It is required that the bolt must withstand shear forces up to 4,000 N. Assuming a safety factor of 5, what must be the minimum diameter of the bolt?*

Solution. Here, the shear force divided by the area of the bolt must be one-fifth of the breaking shear stress. We look up the breaking shear stress for steel in the table and use the value to find

$$\frac{4000 \text{ N}}{\pi d^2/4} = \frac{250 \times 10^6 \text{ Pa}}{5}.$$

Therefore, $d = 10.1$ cm.

5.4 ENERGY IN STRAINED MATERIAL

When an external force causes a strain in a body, the external force does work on the body. This work is stored as potential energy in the strained body. The amount of work done by an external force on deforming a body can be calculated similar to the way we have found the energy stored in a spring when the natural length of the spring is changed. In the case of the spring, we can fix one end of the spring to a rigid support and push or pull on the other end by a force $\vec{F}$. It was best to do the calculation analytically by placing the spring along the x axis with origin at the tip of the spring when the spring is not deformed. Then, deformation of the spring will be given by the x-coordinate of the tip of the deformed spring and the work for changing the length from x to $x + dx$ is easily found to be

$$dW = kxdx, \tag{5.39}$$

since the applied force is matched by the internal force of the spring. Here as usual, k is the spring constant. It is best to write this equation in terms of the deformation dl and the internal force F_s of the spring.

$$dW = -F_s dl, \tag{5.40}$$

where the negative sign means that the force is towards the equilibrium while the deformation is pointed away from the equilibrium. Another thing to note in this formula is that l is the deformation and dl is the change in the deformation. That is we have written the work done by an external agent in changing the deformation of the spring from l to $l + dl$. We can integrate this equation to obtain the energy stored in the spring since we know the law for the force in terms of the deformation l.

$$W_{if} = -\int_i^f F_s dl = k \int_i^f l dl = \frac{1}{2}kl_f^2 - \frac{1}{2}kl_i^2, \tag{5.41}$$

where l_i is the deformation of the spring in the initial state and l_f is the deformation in the final state.

Similarly, we can write the work done for various deformations. Thus, the work done by an external agent in a compressive or tensile strain will be obtained as follows. Let the strain be denoted by s and the stress by the Greek letter σ. Suppose an external force $\vec{F}$ is acting on the body that changes the strain of the body from s to $s + ds$. Since the strain for the extension or compression is defined by the change in length divided by the length, the change in the strain is $ds = dl/l$ and the work written in terms of the strain and the force would be

$$dW = Flds. \tag{5.42}$$

Here l is the length of the object being compressed or stretched. We can convert this to an equation in terms of the stress in the body and the strain of the body by replacing the magnitude of the external force F in terms of the stress in the body.

$$dW = -Al\sigma ds. \tag{5.43}$$

Now, the stress of this type is equal to the product of the Young's modulus Y and the strain s. Furthermore, the direction of the stress must be in the opposite direction to the change in the strain since the stress opposes any change in the strain. Therefore,

$$dW = VYsds, \tag{5.44}$$

where I have replaced Al by the volume V of the object. The energy stored in changing the strain from zero strain to some strain s can be obtained by integrating this equation. This would be the potential energy U stored in the state characterized by the strain value s.

$$U = \frac{1}{2}(VY)s^2, \quad \text{(Compressive and Tensile)}. \tag{5.45}$$

We see that the product VY plays the role of "spring constant" in compressive and tensile strains. Similarly, the work done for shear strain can be written as

$$U = \frac{1}{2}(VG)\theta^2, \quad \text{(Shear)}. \tag{5.46}$$

Finally, we will obtain the following expression for the work done on the body by an external force $\vec{F}$ when the volume changes from V to $V + dV$ with the pressure in the body equal to p.

$$dW = -pdV. \tag{5.47}$$

This work will increase the energy of the body if the body is compressed and decrease the energy if the body expands. The energy stored when the volume of the material changes from the equilibrium value to another value due to applied force on the body can be worked similarly.

5.5 EXERCISES

Static equilibrium

Ex 5.1. A diving board of mass 100 kg and length 4 meters is bolted to two supports separated by 1 meter, one of the support being at the end. A diver of mass 50 kg is standing at the other end of the board. Find the forces at the two support.

Ex 5.2. A sign of mass 300 kg is hanging 1 m from the end of a horizontal beam of mass 50 kg and length 3 m held in place by a pin and a cable as shown in the Fig. 5.14. The cable makes an angle of 30° with the beam. Find the tension in the cable and the net force applied by the pin.

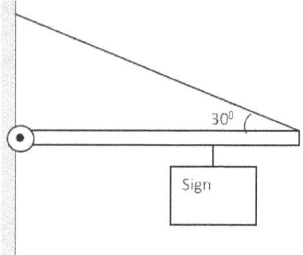

Fig. 5.14: Exercise 5.2.

Ex 5.3. A beam of mass 200 kg and length 3 m is hanging from two cables. One of the cable is attached at the end and the other to the middle of the beam. Find the tension in each cable.

Ex 5.4. A traffic signal of mass 20 kg is suspended over a road by the structure shown in Fig. 5.15. The entire structure is secured by the four bolts at the base. What is the tension in the bolts?

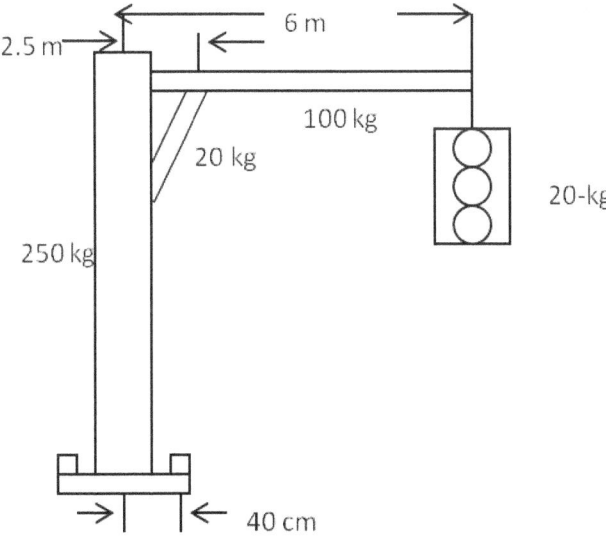

Fig. 5.15: Exercise 5.4

Stress

Ex 5.5. A 30 kg mass is hung from a 1-mm diameter steel wire. Find the stress in the wire and the percent elongation.

Ex 5.6. How much mass must be hung from a 1-mm diameter steel wire before it breaks?

Ex 5.7. While deep sea diving, one must take into account the enormous pressure at great depths under water. At 3000 m depth, the pressure would be approximately 30 MPa, which will crush a ball of steel. (a) What is the percentage change in volume of a spherical solid ball of steel when it is sent to a depth of 3000 m in water? (b) What is the percentage change in diameter?

Ex 5.8. What is the tallest a building one can build using concrete wall of thickness 10 cm? Assume 2400 kg/m^3 for the density of concrete.

Ex 5.9. A rectangular parallelepiped shaped jello with dimensions 5 cm × 4 cm × 3 cm is resting on a plate. A glass slide is glued to the top surface. A force of 30 mN is applied to the glass plate deforming the jello sideways. What is the angle of deformation if the jello has a shear modulus of 1000 N/m^2?

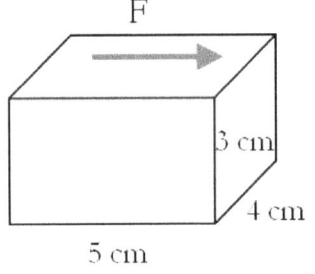

Fig. 5.16: Exercise 5.9.

5.6 PROBLEMS

P 5.1. Rod AB of mass 30 kg and length 1 m is supported by a pin at A and rests on a frictionless peg at C at a distance of 70 cm from A. A force of 1000 N is applied at B as shown. Find the forces on the rod at A and C.

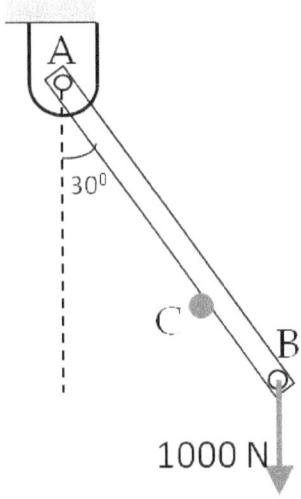

Fig. 5.17: Problem 5.1.

P 5.2. A vertical force is applied at the end B of the bar AB of mass M and length L that is hinged at A and supported by a steel cable of cross-sectional area A attached to the wall at C. Find the percent change in length of the cable if the Young's modulus of steel is Y, distance AC is $\frac{3}{4}L$, and the angle $\angle CAB$ equal θ.

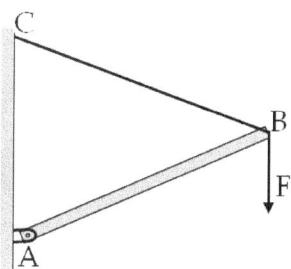

Fig. 5.18: Problem 5.2.

P 5.3. A thin rod of mass M and length L rests on a cylinder of radius R. The rod is attached to a collar of negligible mass, which is free to slide without friction over a vertical guide. Find angle θ at which the rod must rest so that the collar does not slide.

P 5.4. A rod of mass M and length L is supported by three wires as shown. Find tension in each wire if you can. If you cannot find the tension in each wire, then find the tension in the two wires as a ratio of the tension in the third.

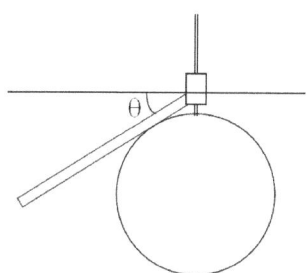

Fig. 5.19: Problem 5.3

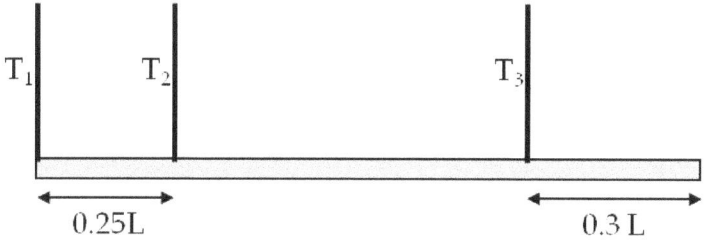

Fig. 5.20: Problem 5.4

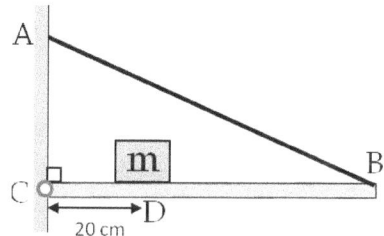

Fig. 5.21: Problem 5.5.

P 5.5. A steel cable AB of diameter 1 mm supports a beam CB of mass 2 kg and length 60 cm hinged at C as shown in the Figure. Assuming the length of the cable to be 65 cm, find the tension in the cable and the force on the pin if mass m = 200 kg.

P 5.6. A 200-kg motorcycle is parked on a slope facing uphill. As a result the normal force on the rear wheel is greater than on the front wheel. If the slope is greater than 40°, the motorcycle slides on the slope. The distance between the front and back wheels is 1.2 meters and the CM of the motorcycle is halfway between the two wheels and 30 cm from the ground as measured in the vertical direction from the slope. The coefficients of static friction between the wheels and the ground are same for both wheels. Find the normal forces on front and rear wheels when it parked at a slope of 40°.

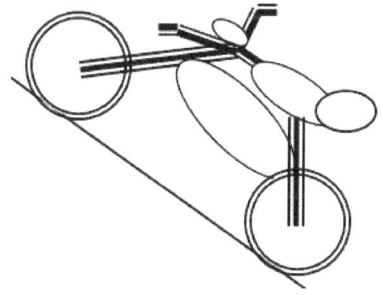

Fig. 5.22: Problem 5.6.

P 5.7. A board of mass m and length L is resting against a vertical wall. The coefficient of static friction between the board and the wall and between the board and the floor are μ_w and μ_f, respectively. Find the minimum angle θ the board may be placed without slipping.

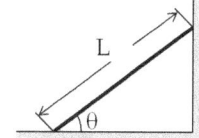

Fig. 5.23: Problem 5.7.

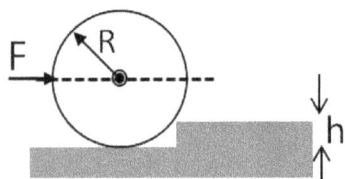

Fig. 5.24: Problem 5.8.

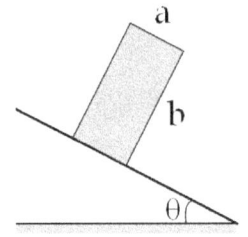

Fig. 5.25: Problem 5.9.

P 5.8. A spherical ball of mass M and radius R is resting against a step of height h ($h<R$). A horizontal force $\vec{F}$ is applied to the ball so that its line of action always passes through the center of the ball, as shown in the figure. Find the minimum force needed to make the ball climb over the step.

P 5.9. A rectangular parallelepiped shaped wooden block of height b and square base of side a is placed on smooth plane. There is a static friction between the block and the plane. As the angle of the plane is varied, the block may either tip over first or slide first depending upon the values of a, b and θ. Find the condition when the block tips first.

Chapter 6

FLUID STATICS

Contents

Liquid and gases are collectively called fluids. Fluids differ from solids in their response to a force applied tangential to the surface, also called a shear force. Fluids flow when a shear force is applied on them, but solids do not; a solid may be temporarily deformed, but when the shear force is removed a solid tends to return to its original shape if the shear stress is not too great. The microscopic reason for the difference is in the rigidity of intermolecular bonding in solids compared to liquids and gases. The molecules of a fluid are able to move around relatively easily, while in solids, they are fixed in their places.

Although liquids and gases have similarities in their flow properties, there is one important difference: gases fill up the entire space of the container regardless of the amount of gas in the container while a liquid fills only part of the container depending upon the amount of the liquid. Due to gravity, a liquid occupies the lowest part of a container if the flow is not constricted. Liquids and solids are also much less compressible than a gas. Despite their differences, it is useful to treat liquids and gases together as far as their flow properties are concerned. In this chapter, we will study physical properties of static fluids, such as pressure, density and volume, and in the next chapter we will look at the dynamics of flow.

6.1 DENSITY AND INCOMPRESSIBLE FLUIDS

Density (ρ) of a material is defined as mass (M) divided by volume (V).

$$\rho = \frac{M}{V} \tag{6.1}$$

Density has the dimension of mass over length cubed, and has units of kg/m^3 in kg-m-sec system of the SI units. In the cgs system, the unit for density is g/cc, or gram per cubic cm, which is equal to one-thousand kg/m^3.

$$1 \text{ g/cc} = 1000 \text{ kg/m}^3.$$

The density of a homogeneous material is same everywhere, but for an inhomogeneous material, such as a mixture of two immiscible substances, the density will depend on the location. Therefore, we will introduce a concept, called the local density, $\rho(x, y, z)$, which is a function of the coordinates of the space point (Fig. 6.1).

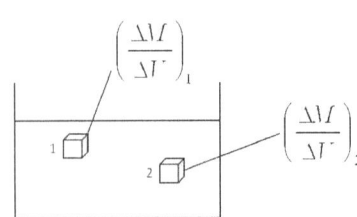

Fig. 6.1: Density may be different at different places. Local density at a point is obtained from mass over volume in small volume around that point.

Local density can be obtained by a limiting process from the average density in a small volume around the space point. We write this process as a limit.

$$\rho(x, y, z) = \lim_{\Delta V \to 0} \left(\frac{\Delta M}{\Delta V} \right) \qquad (6.2)$$

The densities of gases vary considerably with temperature, while the densities of liquids vary little with temperature. Therefore, we will treat the densities of liquids as constant. The variation of the density of gases with temperature will be studied in the chapter on heat and temperature.

The density is a dimensionful property. Therefore, you need to keep track of units to compare two densities. For comparison purposes a more convenient dimensionless quantity called the **specific gravity** is constructed. Specific gravity is defined as the ratio of the density of the material to the density of water at 0°C and one atmospheric pressure, which is 1 g/cc.

$$\text{Specific gravity} = \frac{\text{Density of material}}{\text{Density of water}}$$

Specific gravity, being dimensionless, provides a ready comparison among materials without having to worry about the unit of density. For instance density of aluminum is 2.7 in g/cc and 2700 in kg/m^3, but specific gravity is 2.7 regardless of the unit of density. The density of some common substances are listed in Table 6.1.

We will be dealing with fluids whose density does not change appreciably regardless of the physical conditions imposed them. We call these fluids incompressible. Liquids are to a large extent incompressible. Gases are quite compressible as you can easily demonstrate by filling a container with a gas. If the container is made up of a rigid material and the lid of the container can slide, you will find that the lid can be moved in and out to change the volume occupied by the same gas. This makes it easy to change the density of a gas.

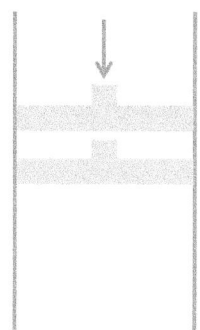

Fig. 6.2: Gases are much more compressible than liquids or solids.

6.2 PRESSURE

The contact forces such as the normal force, frictional force, push, etc, act at the surface of a body. An important characteristics of fluids is that there is no significant reaction force from the fluid when you apply a force horizontal to the surface of the fluid. The fluid simply flows under such a force.

Table 6.1: Density of some common substances at 0°C and 1 atm in g/cc. To obtain density in kg/m^3 multiply by 1000.

Material	Density (g/cc)	Material	Density (g/cc)
Solids		**Liquids**	
Aluminum	2.7	Alcohol, ethanol	0.81
Brass	8.6	Blood	1.05
Concrete	2	Gasoline	0.68
Copper	8.9	Glycerol	1.26
Glass	2.5	Mercury	13.6
Granite	2.7	Olive oil	0.92
Gold	19.3	Sea water	1.03
Ice	0.92	Water (40°C)	1.0
Iron	7.8	**Gases**	
Lead	11.3	Air	1.3×10^{-3}
Platinum	21.4	Carbon dioxide	2.0×10^{-3}
Silver	10.5	Helium	2.0×10^{-4}
Steel	7.8	Water (*steam*, 100°C)	6.0×10^{-4}
Wood	0.5		

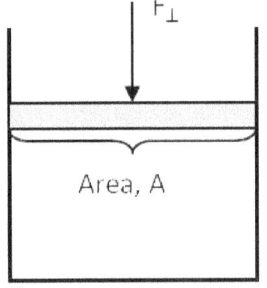

Fig. 6.3: Defining pressure. Pressure is equal to the perpendicular component of force divided by area over which the force acts.

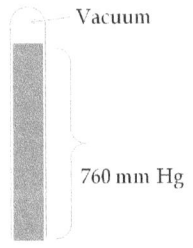

Fig. 6.4: Atmospheric pressure can support the weight of a 760 mm column of mercury.

A vertical force, however, compresses or expands the fluid. Suppose you try to compress a fluid, then you would find that a reaction force develops at each point inside the fluid in the outward direction that balances the force applied on the molecules at the boundary. **Pressure** (p) is defined to be the normal force ($F_\perp$) divided by the area (A) over which the force acts (Fig. 6.3):

$$p = F_\perp/A. \tag{6.3}$$

The dimension of pressure is

$$[M][L]^{-1}[T]^{-2},$$

and therefore the unit in SI system is kg/m.s^2 or N/m^2, which is named **Pascal (Pa)** in honor of the French mathematician and physicist Blaise Pascal (1623-1662). Actually, other units of pressure are in more common use in physics and engineering. For instance, the ambient pressure due to the atmosphere is called **one atmosphere (atm)** which is approximately 1.013×10^5 Pa. We will see below that the pressure at the bottom of a 760-mm high column of mercury at 0°C in a container where the top part is evacuated is equal to the atmospheric pressure. Thus, 760-mmHg is also used in place of 1 atmospheric pressure.

In vacuum physics labs, one often uses another unit called Torr, named after the Italian inventor Evangelista Torricelli (1608-1647), who invented the mercury manometer for measuring pressure. One Torr is equal to a pressure of 1-mmHg. Low pressure in SI units is sometimes expressed in yet another unit called millibar or mbar, which is equal to 100 Pa, with a bar being 10^5 Pa. The multitude of units for pressure is certainly very confusing to students. A summary of their relations is provided in the margin for a quick reference.

The inverse area in the definition of pressure has interesting physical implications. For instance, a force applied on an area of 1 mm^2 has a pressure that is 100 times as great as the same force applied on an area of 1 cm^2. That is why a sharp pin is able to poke though skin when a small force is exerted while a blunted metal rod cannot with the same force. When a force is applied to the pin, the pressure applied on the skin is balanced up to a maximum. But, since the area of the tip of the pin is small, the pressure from the applied force is great. Pressure form the outside agent on the area of the skin at the tip far exceeds the pressure on the skin from inside the body and that applied by adjoining parts of the skin. This makes it possible for a sharp needle to pierce the skin more easily than a dull needle.

Many units of pressure
SI unit:
1 Pa = 1 N/m^2
English unit:
pounds per square inch (lbs/in^2 or psi),
1 psi = 7.015×10^3 Pa

Other units of pressure:
1 atm = 760 mmHg = 1.013 $\times 10^5$ Pa = 14.7 psi = 29.9 inches of Hg
1 Torr = 1 mmHg = 133.29 Pa
1 bar = 10^5 Pa

Table 6.2: Pressures in universe (Source: various)

Place	Pressure (Pa)	Place	Pressure (Pa)
Center of the Sun	$\approx 3x10^{16}$	Mechanical Vacuum Pump	100 to 10^{-4}
Center of the Earth	$\approx 4x1011$	Cryopump molecular beam epitaxy	10^{-9}
Atmosphere	1.013×10^5	On the Moon	10^{-9}
Vacuum cleaner	$\approx 8 \times 10^4$	Interstellar space	10^{-15}

6.3 PRESSURE IN A STATIC FLUID NEAR EARTH

Pressure in a fluid near earth varies with depth due to the different weight of fluid above a particular level. We can understand this by imagining a surface at a given level and noting that at the deeper level, more weight of the fluid acts on the imagined surface. Therefore we expect a greater pressure at the deeper level. Note that if the same fluid is taken in outer space of zero gravity then this effect will not be present any more.

In this section we will derive a formula for the variation of pressure with depth in a tank containing a fluid of density ρ on the surface of Earth. Imagine a thin element of fluid at a depth h. Let the element have a cross-sectional area A and height Δy. Then, the forces on the element are from the pressures $p(y)$ above and $p(y+\Delta y)$ below it, and the weight of the element itself as shown in the free-body diagram in Fig. 6.5.

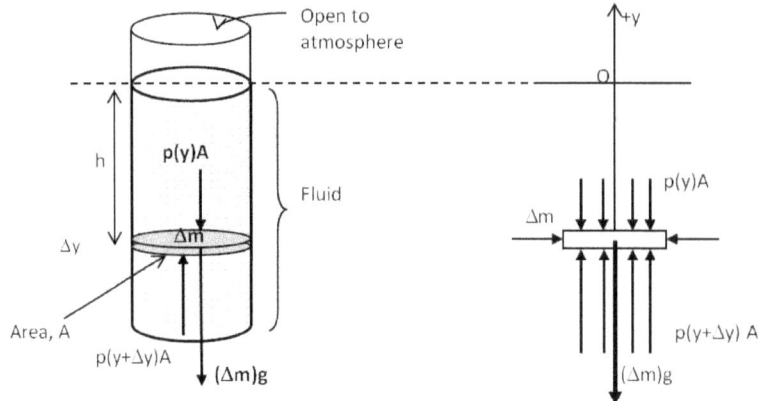

Fig. 6.5: Forces on a mass element inside a fluid.

Since the element of fluid between y and $y+\Delta y$ is not accelerating, the forces are balanced. Using the Cartesian y-axis pointed up, we find the following equation for the y-component.

$$p(y + \Delta y) - p(y)A - g\Delta m = 0 \quad (\Delta y < 0) \qquad (6.4)$$

Note that if the element had a non-zero y-component of the acceleration, the right side would not be zero, it would be the mass times the y-acceleration instead. The mass of the element can be written in terms of the density of the fluid and the volume of the elements.

$$\Delta m = |\rho A \Delta y| = -\rho A \Delta y \quad (\Delta y < 0). \qquad (6.5)$$

Putting the expression for Δm from Eq. 6.5 into q. 6.4, and then dividing both sides by $A\Delta y$ we find the following.

$$\frac{p(y + \Delta y) - p(y)}{\Delta y} = -\rho g \qquad (6.6)$$

Taking the limit of infinitesimally thin element $\Delta y \to 0$, we discover the following differential equation that gives the variation of pressure in a fluid.

$$\boxed{\frac{dp}{dy} = -\rho g.} \qquad (6.7)$$

This is an important equation that tells us that the rate of change of pressure in a fluid is proportional to the density of the fluid. The solution of this equation will depend upon whether the density ρ is constant or changes with depth, i.e. the function $\rho(y)$.

To be sure, if the range of the depth being analyzed is not too great, we can assume the density to be constant. But, if the range of depth is great, such as in the case of the atmosphere, there would be significant change in density with depth. In that case, we cannot use the approximation of a constant density. In our first application of Eq. 6.7, we will work out a formula for the pressure in a tank of liquid such as water, where the density of liquid can be taken to be constant.

Example 6.1. Constant density fluid

We now consider a simple application of Eq. 6.7 and find a formula for the pressure at a depth h from the surface in a fluid of constant density. We need to integrate the equation from $y = 0$, where the pressure is the atmospheric pressure (p_0), to $y = -h$, the y-coordinate of the depth.

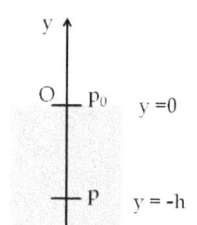

Fig. 6.6: Example 6.1.

$$\int_{p_0}^{p} dp = -\int_{0}^{-h} \rho g \, dy \implies p - p_0 = \rho g h. \qquad (6.8)$$

Hence pressure at a depth of a fluid on the surface of earth is equal to the atmospheric pressure plus $\rho g h$ if the density of the fluid is constant over the height.

Note that the pressure in a fluid only depends on the depth from the surface and not on the shape of the container. Therefore if you construct a container where a fluid can freely move in various parts, the liquid will be at the same level in every part regardless of the shape as shown in Fig. 6.7.

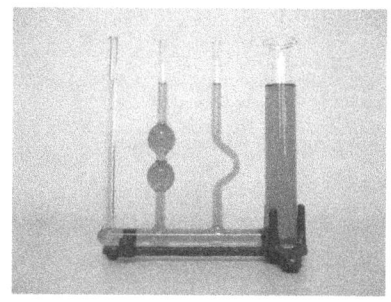

Fig. 6.7: If the fluid can flow freely, it rises to the same height in each part.

Example 6.2. Variation of Atmospheric Pressure With Height

The change in atmospheric pressure with height is of particular interest. Assuming the temperature of air to be constant, and that ideal gas law of thermodynamics describes the atmosphere to a good approximation, find the variation of atmospheric pressure with height. [Note added: We will study the ideal gas law in more detail in a later chapter. Here we just make use of that law for illustrative purposes.]

Solution. Let $p(y)$ be the atmospheric pressure at height y. The density ρ at y, the temperature T in the Kelvin scale (K), and the mass m of a molecule of air are related to the absolute pressure by the ideal gas law given as

$$p = \rho \frac{k_B T}{m} \quad \text{(atmosphere)} \qquad (6.9)$$

where k_B is a constant called Boltzmann's constant that has a value of 1.38×10^{-23} J/K. You may have studied the ideal gas law in the "chemistry" notation, $PV = nRT$, where n is the number of moles, and R the gas constant. Here, the same law has been written in a different form, using the density ρ instead of volume V. Therefore, if pressure p changes with height so would the density ρ. Using density from the ideal gas law, we find that rate of variation of pressure with height is given as

$$\frac{dp}{dy} = -\left(\frac{mg}{k_B T}\right) p, \qquad (6.10)$$

where constant quantities have been collected inside the parenthesis. Replacing the quantity within the parenthesis () by a single symbol α the equation looks much simpler.

$$\frac{dp}{dy} = -\alpha p. \qquad (6.11)$$

Now, we can solve this equation by noticing that the pressure p is a function whose derivative is proportional to itself. Since the exponential functions are such functions we try an exponential function and sure enough find the solution immediately.

$$p(y) = p_0 \exp(-\alpha y) \qquad (6.12)$$

Thus atmospheric pressure drops exponentially with the height since the y-axis is pointed up from the ground and y has positive values in the atmosphere above sea level. The pressure drops by a factor of $\frac{1}{e}$ when the height is $\frac{1}{\alpha}$, which gives us a physical interpretation for α: constant $\frac{1}{\alpha}$ is a length scale that characterizes how pressure varies with height.

An approximate value of α can be obtained by using mass of a nitrogen molecule as a proxy for an air molecule. At temperature $27°C$ or $300K$, we find

$$\alpha = -\frac{mg}{k_B T} = \frac{4.8 \times 10^{-26} \text{ kg} \times 9.81 \text{ m/s}^2}{1.38 \times 10^{-23} \text{ J/K} \times 300 \text{ K}} = \frac{1}{8,800 \text{ m}}. \qquad (6.13)$$

Therefore, for every 8,800 meters, the air pressure drops by a factor $\frac{1}{e}$ or approximately one-third of its value. Of course, this gives us only a rough estimate of the physical situation since we have assumed both the temperature and g constant over such great distances from the Earth, neither of which is correct in reality.

6.4 PRESSURE MEASUREMENTS

The variation of pressure with height provides a simple way to measure pressure. This principle was utilized in the invention of mercury barometer by the Italian mathematician and physicist Evangelista Torricelli (1608-1647) in 1643. Since then, many other techniques of pressure measurement have been developed.

Any property that changes with pressure in a known way can be used to construct pressure-measuring devices called **pressure gauges**. The most common ones are the strain gauges which use the change in shape of a material with pressure, the capacitance pressure gauges which use the change in capacitance due to shape change with pressure, the piezoelectric pressure gauge which generates a voltage difference across a piezoelectric material under pressure difference between the two sides, and the ion gauges which measures pressure by ionizing molecules in highly evacuated chambers. Different pressure gauges are useful in different pressure ranges and under different physical situations. Here I will describe only two instruments, namely the barometer and a related instrument called the manometer.

Barometer

A mercury barometer is a tube with an opening at one end. The tube is filled with mercury and the open end covered so that no air is trapped in the tube when the tube is inverted. The filled tube is then inverted and opened in a mercury tank such that no air can enter the tube. Torricelli found that mercury in the tube empties out to a height of approximately 29.9 inches. Originally, when the same experiment was done with water in the inverted tube, it was found that the tube was filled with water to increasing heights in longer and longer tubes. As a matter of fact atmospheric pressure can support approximately 34 feet of water column. It was interpreted to indicate that nature "abhors" vacuum. Torricelli was Galileo's secretary in the last three months of latter's life. Galileo apparently suggested to Torricelli to try mercury since mercury was a much denser substance and a smaller tube might be sufficient as found later by Torricelli. Torricelli correctly interpreted that the empty space above mercury in the tube was vacuum, and in this way he was able to create sustained vacuum for the first time in science.

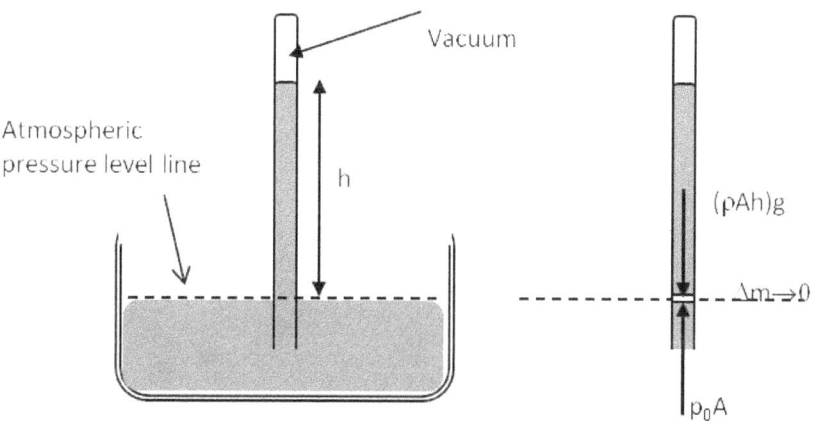

Fig. 6.8: Torricelli§s experiment.

Using modern analysis we can understand how the height of mercury in the tube is directly related to the atmospheric pressure. Recall that the pressure at the same horizontal level in one fluid is same whether in the tube or outside as long as the fluid can flow in and out of the tube freely. Consider an infinitesimally small element of mercury at the level point (Fig. 6.8). Since the mass element is not accelerating the net force on the element must be zero. From the free-body diagram the equation for vertical direction is found to be

$$p_0 A - \rho A h g - \Delta m g = 0 \quad (\Delta m \to 0) \tag{6.14}$$

With $\Delta m \to 0$, we find the following for atmospheric pressure p_0.

$$p_0 = \rho g h. \tag{6.15}$$

Hence, the height of the liquid in the tube is directly proportional to the atmospheric pressure in the barometer. Our analysis also shows that higher the density of the fluid, the lower the height h for the same pressure p_0. Since water is approximately 13.6 times less dense, you will need $13.6 \times 29.9 = 407$ inches tall tube to measure atmospheric pressure if you use water in place of mercury!

Closed-tube Manometer

The pressure of a gas can be measured by making a modification to the barometer. The inverted tube of Torricelli's experiment is made into a U-shaped tube with the open end connected to the gas container whose pressure we wish to measure (Fig. 6.9). The instrument is called a **manometer**.

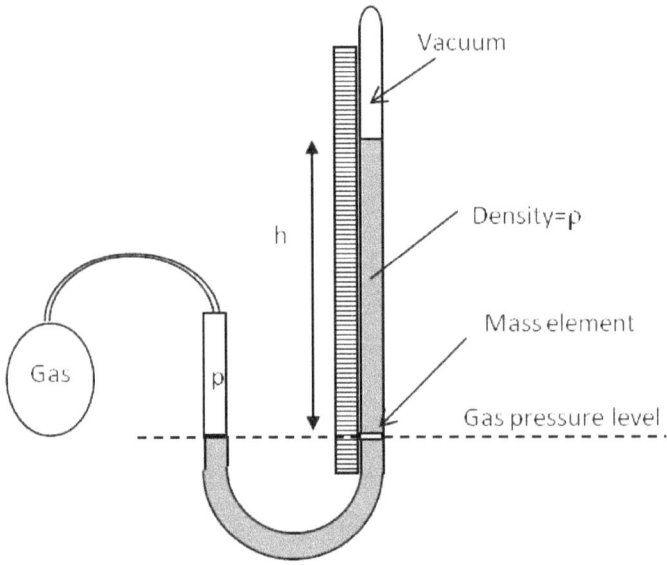

Fig. 6.9: Manometer.

Applying Newton's second law of motion on an infinitesimal fluid element Δm leads to the following equation.

$$pA - \rho Ahg - \Delta mg = 0 \quad (\Delta m \to 0) \tag{6.16}$$

where p is the pressure of the gas, A the area of the cross-section of the tube, h the height difference on the two sides of the U-tube and ρ the density of the fluid in the tube (not the density of gas). Simplifying the equation we find the following for the pressure reading for the gas.

$$p = \rho gh. \tag{6.17}$$

The closed end of a manometer are calibrated and marked for directly reading off the pressure.

Open-end manometer

Manometers can also be constructed to read the difference of pressure of the gas and the atmospheric pressure, called the gauge pressure, if both ends of the U-tube are open as I will show now. One end of the U-tube is opened to air and the other to the gas as shown in the Fig. 6.10.

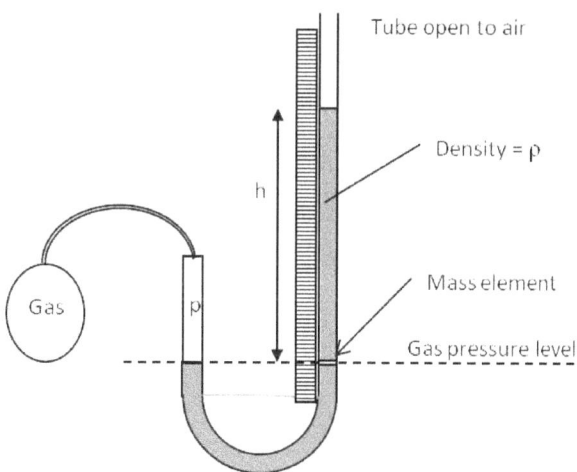

Fig. 6.10: Open-end Manometer.

The competition between the pressures of gas and atmosphere results in the difference in height of the liquid in the U-tube. The height difference of the liquid in the two arms of the U-tube is the **gauge pressure** of the gas. If the gas pressure is higher than atmospheric pressure then the level of fluid in the gas side will be lower than the one exposed to the atmosphere, while if the gas pressure is less than the atmospheric pressure then the level on the gas side will be higher.

Apply Newton's second law of motion on an infinitesimal fluid element at the lower fluid level in the arm with higher fluid level.

Fig. 6.11: Gas tank pressure gauges.

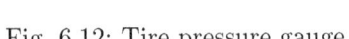

Fig. 6.12: Tire pressure gauge.

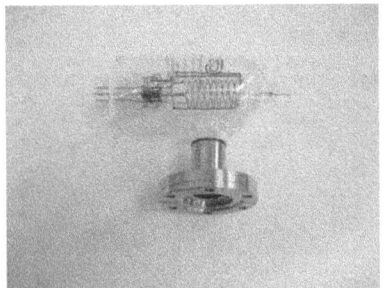

Fig. 6.13: Ionization gauge.

In the figure, the fluid element is on the atmosphere side. Let p be the pressure of the gas, p_0 the atmospheric pressure, A the area of the cross-section of the tube, h the height the fluid and ρ its density (Fig. 6.10). Free-body diagram for the forces on the infinitesimal mass element leads to the following equation.

$$pA - (\rho h g + p_0)A - \Delta m g = 0 \quad (\Delta m \to 0) \qquad (6.18)$$

Simplifying we get the pressure of the gas to be

$$p = p_0 + \rho g h \qquad (6.19)$$

The difference of actual or absolute pressure p and the atmospheric pressure p_0 is also called the gauge pressure.

$$p_0 = \text{Atmospheric pressure}$$
$$p = \text{Absolute pressure}$$
$$p - p_0 = \text{Gauge pressure}$$

Pressure gauges often measure the difference of the absolute pressure and the ambient atmospheric pressure and hence the name gauge pressure. Open manometer is one kind of pressure gauge. Many mechanical and electrical pressure gauges are available commercially which are more convenient to use than the open manometer (see margin for examples of pressure gauges).

Example 6.3. Fluid Heights in an Open U-tube.

A U-tube with both ends open is filled a liquid of density ρ_1 to a height h on both sides. A liquid of density $rho_2 < \rho_1$ is poured on one side. We find that the liquid 2 settles on top of the liquid. The heights on the two sides are different. The height to the top of the liquid 2 from the interface is h_2 and the height to the top of liquid from the level of the interface is h_1. Deduce a formula for the height difference.

Solution. The pressure at the same height on the two sides of a U-tube must be same as long as the two points are in the same liquid. Therefore, we consider two points at the same level in the two arms of the tube: One point will be the interface on the side of the liquid 2 and the other will be a point in the arm with liquid 1 that is at the same level as the interface in the other arm.

Pressure on the side with liquid 1 $= p_0 + \rho_1 g h_1$.

Pressure on the side with liquid 2 $= p_0 + \rho_2 g h_2$.

Since, the two points are in the liquid 1 and are at the same height, the pressure at the two points must be the same pressure. Therefore, we have

$$p_0 + \rho_1 g h_1 = p_0 + \rho_2 g h_2.$$

Hence,

$$\rho_1 h_1 = \rho_2 h_2.$$

This says that the difference in heights of the two sides on the U-tube would be

$$h_2 - h_1 = \left(1 - \frac{\rho_1}{\rho_2}\right) h_2.$$

The result makes sense if we set $\rho_2 = \rho_1$, which gives $h_2 = h_1$. If the two sides have the same density, they would have the same height.

6.5 FLUIDS AT REST AND PASCAL'S PRINCIPLES

When a fluid is not flowing, we say that the fluid is in a static equilibrium. If the fluid is water the term **hydrostatic equilibrium** is used. In this condition, the net force on any part of a fluid at rest must be zero, otherwise the fluid will start to flow.

In 1653 the French mathematician, and physicist Blaise Pascal (1623-1662) published his Treatise on the Equilibrium of Liquids in which he outlined three principles of static fluids.

6.5.1 Pascal's First Principle

The force at any point inside a fluid due to the pressure at that point acts with the same magnitude in all directions.

That is, the force due to the pressure in a fluid will be the same on any area at an infinitesimal particle oriented in whichever manner (Fig. 6.14).

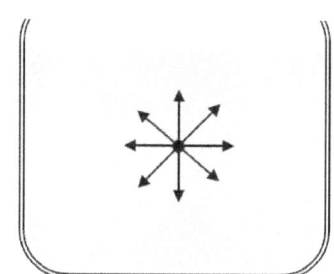

Fig. 6.14: Pressure at a point is isotropic, meaning same in all directions from the point.

6.5.2 Pascal's Second Principle or Simply Pascal's Principle

When pressure is changed (increased or decreased) at any point in an incompressible fluid, the pressure at all other points in the fluid changes to the same extent.

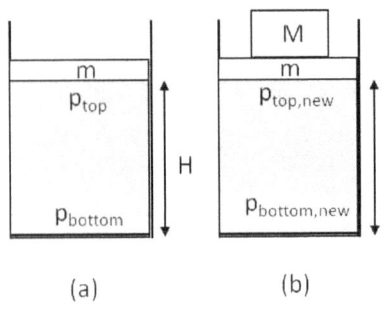

Fig. 6.15: Pressure in fluid is changed when the fluid is pressed. The pressure at the top layer of the fluid is different than at the bottom layer. The increase in pressure by putting additional weight is same everywhere, e.g., $p_{\text{top new}} - p_{\text{top}} = p_{\text{bottom new}} - p_{\text{bottom}}$.

The key phrase here is "change in pressure". Often, the second principle of Pascal is also referred to as the **Pascal's principle**. Note that this principle does not say that the pressure is same at all points of a fluid, which is certainly not correct since the pressure in a fluid near earth varies with height. This principle rather addresses the change in pressure.

Suppose you have some water in a cylindrical container of height H and cross-sectional area A that has a move-able piston of mass m (Fig. 6.15). Adding weight Mg at the top slide-able lid would increase the pressure at the top by Mg/A, since the additional weight also acts over area A of the lid.

$$\Delta p_{\text{top}} = \frac{Mg}{A}$$

According to Pascal's second principle, the pressure at all points of water will change by the same amount Mg/A. Thus, the change in pressure at the bottom will also be an increase of Mg/A. This seems surprising since one may expect the pressure at the top to change more, but that is not the case.

$$\Delta p_{\text{bottom}} = \frac{Mg}{A}$$

Since the pressure changes are the same everywhere in the fluid, we do not need to label the change with subscripts top or bottom and will drop the subscripts.

$$\Delta p = \Delta p_{\text{top}} = \Delta p_{\text{bottom}} = \Delta p_{\text{everywhere}}$$

6.5.3 Pascal's Third Principle

In a hydraulic jack, a small force applied over a small area can balance a much larger force on the other side over a larger area.

A **hydraulic jack** is a device that has an incompressible fluid in a U-tube fitted with move-able pistons on the two sides. One side of the U-tube is narrower than the other to make use of Pascal's third principle (Fig. 6.16).

When a force $\vec{F}_1$ is applied on the side with the area of cross-section A_1 it adds pressure P_1 to all points of the fluid according to the second principle stated above. In particular an increase of pressure P_1 in the upward direction takes place on the piston A_2 on the other side. A second force must be applied on the piston A_2 to balance the effect of the increased pressure. Let $\vec{F}_2$ be the force on the second piston with the area of cross-section A_2 to balance the force on the second

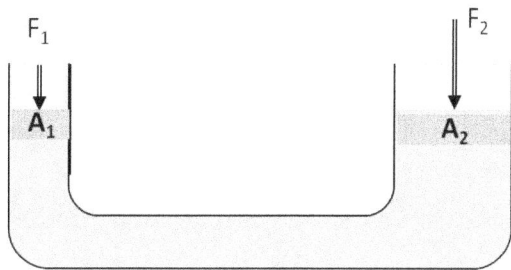

Fig. 6.16: A hydraulic jack.

piston. Then we must have

$$\frac{F_1}{A_1} = \frac{F_2}{A_2}.$$ (6.20)

Example 6.4. Lifting a car

Find the force needed to balance a 3000-kg car on the wider side of a hydraulic jack with the circular cross-sections of radii 10 cm and 0.5 cm.

Solution. From Pascal's third principle, it is immediately evident that the following force will be needed to balance the weight of the car.

$$F_1 = \frac{A_1}{A_2} F_2 = \left(\frac{\pi r_1^2}{\pi r_2^2} \right) F_2 = 73.5 \ N,$$

which is the weight of a 7.5 kg mass.

6.6 THE ARCHIMEDES'S PRINCIPLE

It is said that once King Hiero of Syracuse, a port city in Greece, ordered a gold crown and gave the exact amount to a goldsmith. When the king receive the crown, it had the correct weight, but he suspected that some of the gold was replaced by some other metal. The king asked the famous geometer of the time, Archimedes (287? BC 3 212 BC), to find out if the crown was made of gold without destroying it. This was a difficult problem at the time and no one knew how to do it.

The idea came to Archimedes one day as he was entering his bathtub. He noticed that the amount of water that overflowed the tub was proportional to the amount of his body submerged. He also recognized that he could use this to solve the mystery of the crown. He was very excited by this discovery and ran through the streets of Syracuse shouting "Eureka! Eureka!" (I have found it!).

Archimedes's principle refers to the force of buoyancy that results when a body is submerged in a fluid, whether partially or wholly. The pressure of a fluid acts on a body perpendicular to the surface of the body, always pointed into the body. That means that the pressure at the bottom part is pointed up while at the top part is pointed down and the pressures at the sides are pointing into the body.

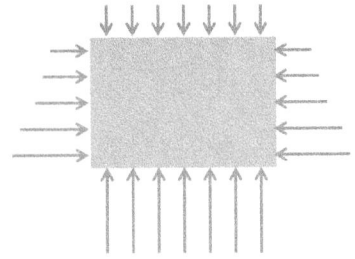

Fig. 6.17: The cause of buoyancy is the greater pressure at the higher depth. the higher pressure at the lower part of the body gives a net upward force to the body.

Since the bottom of the body is at a greater depth than the top of the body, the pressure at the lower part of the body would be higher than the pressure at the upper part as shown in Fig. 6.17. Therefore, there will be a net upward force on the body. This upward force is called the **force of buoyancy** or simply the **Buoyancy**. Archimedes found a quantitative rule for the magnitude of the buoyancy force, which is codified in the **Archimedes's principle**.

The magnitude of the buoyancy force is equal to the weight of the displaced fluid and the direction of buoyancy force is opposite to the direction of the weight.

Let ρ_0 be the density of the fluid and V_{sub} the volume of a material submerged in the fluid, then according to Archimedes's principle, the force of buoyancy has the following magnitude and direction.

$$\text{Magnitude: } F_B = m_{\mathrm{disp}}g = (V_{\mathrm{sub}}\,\rho_0)\,g, \quad \text{Direction: } \uparrow \qquad (6.21)$$

where m_{disp} is the mass of the displaced fluid. For the purposes of calculating torque on the body due to the force of buoyancy, the buoyancy force can be thought to act at the center of buoyancy (CB), which coincides with the center of mass (CM) of the displaced fluid (Fig. 6.18) and not necessarily at the CM of the body if the body is only partially submerged. If the body is fully submerged, the CM of the body and CB coincide.

6.6.1 Using Archimedes's Principle to Find Density

Consider a material of an unknown density ρ. Let the mass of the sample be M. When the sample is dipped in a fluid of a known density ρ_0 its apparent weight is less than it weight Mg. Let us consider the case of $\rho > \rho_0$ so that the sample sinks in the fluid and can be fully submerged. The density of the unknown can be found from its apparent weight through the following procedure.

Tie a thread of negligible mass to the sample and hang it to a spring balance. Now, submerge the sample in the given fluid making sure

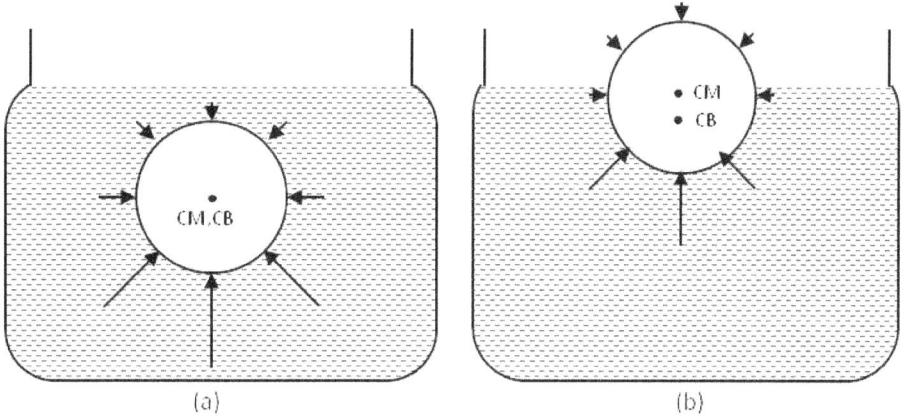

Fig. 6.18: Buoyancy in (a) fully submerged and (b) partially submerged material arises due to the different pressure at different depths in the fluid. The center of buoyancy (CB) is at the center of mass of the displaced fluid which may not be at the center of mass of the body itself.

not to touch the bottom of the container, and record the apparent weight reading W' in the spring balance (Fig. 6.19).

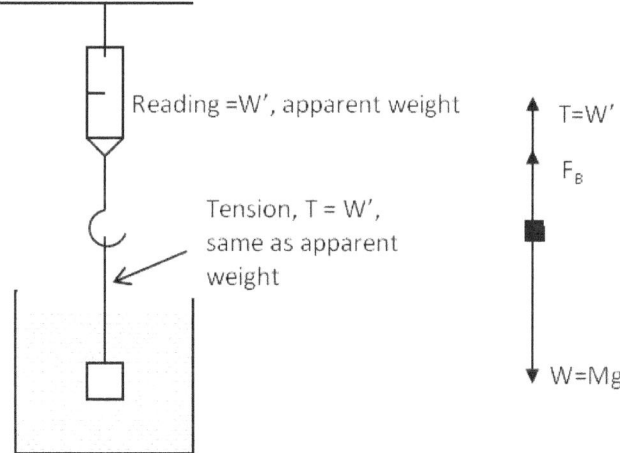

Fig. 6.19: Finding density of a fully submerging substance in a liquid.

The forces on the mass when it is submerged are: weight Mg, buoyancy force F_B and tension $T = W'$, all of whom act in the vertical direction. Since the mass has no acceleration, Newton's second law of motion gives us the following relation for the vertical direction.

$$T + F_B - W = 0 \qquad (6.22)$$

For the forces we have the following expressions.

$$W = Mg$$

$$T = W'$$

$$F_B = V_{\text{submerged}} \times \rho_0 \times g = V_{\text{sample}} \times \rho_0 \times g \quad (\text{since fully submerged})$$

$$= \left(\frac{M}{\rho} \right) \times \rho_0 \times g = \frac{\rho_0}{\rho} \times W$$

After putting these equations in Eq. 6.22 we can solve for the density ρ of the sample in terms of the apparent weight W' while fully submerged in the fluid, weight W, and density ρ_0 of the fluid.

$$\rho = \left(\frac{W}{W - W'} \right) \rho_0 \qquad (6.23)$$

Using this relation you could determine the density of the crown given to Archimedes without knowing the exact volume of the crown. Legend has it that the jeweler did cheat the king and he was beheaded when Archimedes found the cheat.

Example 6.5. Apparent Weight of Steel Ball in Oil.

A 100-g steel ball is submerged in oil of density 0.8 g/cc. Find the apparent weight if the density of steel is 8.0 g/cc.

Solution. Solving for W' in Eq. 6.23, we find the following.

$$W' = \left(1 - \frac{\rho_0}{\rho} \right) W$$

$$= \left(1 - \frac{0.8}{8.0} \right) \times 0.100 \text{ kg} \times 9.81 \text{ m/s}^2$$

$$= 0.882 \text{ N}.$$

The weight is 0.981 N and the apparent weight 0.882 N.

6.6.2 Buoyancy for Partially Submerged Bodies

Case 1 $(\rho > \rho_0)$

A compact object whose density ρ is greater than the density of the fluid ρ_0 will sink in the fluid. To submerge it partially we can hang the object in the fluid by some mechanism. As more of the object is submerged the apparent weight decreases till when all of the body is inside the fluid. After that, the apparent weight does not change. We will find a formula that relates the apparent weight to the fraction of the volume submerged.

The relation between W, W' and F_B given in equation 6.22 still holds. Let x be the fraction of volume of the sample submerged, then the vertical component of forces on the body give the following relation.

$$T + F_B - W = 0$$
$$\implies W' + x\frac{\rho_0}{\rho}W - W = 0$$
$$\implies W' = \left(1 - x\frac{\rho_0}{\rho}\right)W \qquad (6.24)$$

Case 2 $(\rho \le \rho_0)$

An object whose density is less than that of the fluid will float in the fluid. Here, we will find a formula for the fraction of the volume that will be submerged when a body of density less than the density of the fluid floats on the surface of the fluid. You may think of this as an application of the partially submerged case where $W' = 0$ since if you were to use a string to lay the sample on the surface of the fluid there will be zero tension in the string when you let go of it. Setting $W' = 0$ in Eq. 6.23 we find the fraction of volume submerged is given by the ratio of the densities.

$$x = \frac{\rho}{\rho_0} \qquad (6.25)$$

Of course, you can also do this problem directly. Refer to Fig. 6.20 for the free-body-diagram. Since vertical acceleration of the mass is zero, the second law of motion gives us the following relation.

$$xV\rho_0 g - V\rho g = 0 \implies x = \frac{\rho}{\rho_0} \qquad (6.26)$$

Even an object whose density is more than that of fluid can be made

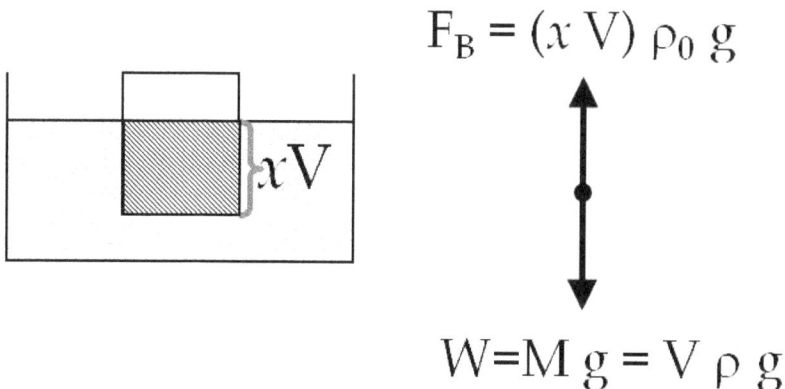

Fig. 6.20: Partially submerged object.

to float in the fluid if it is shaped such that the displaced fluid has more weight than the weight of the object itself. This is the principle by which a boat made of steel floats in water even though the density of steel is approximately eight times that of water. The boat has a shape that lets it displace more volume of fluid than the volume of the steel itself. For a steel boat to float it must displace at least eight times its volume.

Example 6.6. Density of Wood Floating in Water.

While freely floating, sixty five percent of the volume of a block of wood is found to be submerged in salt water of density 1.2 g/cc. What is the density of the wood?

Solution. Since freely floating, we can use $x =$ ratio of densities.

$$\rho = x\rho_0 = 0.65 \times 1.2 \text{ g/cc} = 0.78 \text{ g/cc}.$$

Example 6.7. Partially Submerged Hollow Steel Ball.

A hollow spherical steel ball with inner radius 2 cm and outer radius 2.1 cm floats freely in saltwater of density 1.1 g/cc. What is the percentage of total volume submerged under saltwater? Assume density of steel to be 8 g/cc.

Solution. The hollow steel ball has an effective density much less than the density of compact steel. What matters for buoyancy is the displaced fluid which depends on the outer surface of the sample. Therefore we calculate the effective density of the sample from the mass of the given sample and the volume of the outer surface.

$$M = \frac{4}{3}\pi \left(r_{out}^3 - r_{in}^3 \right) \rho_{steel} = 46.5 \text{ g}.$$
$$V_{out} = 44.6 \text{ cc}.$$

The effective density of the sample is:

$$\rho_{eff} = \frac{M}{V_{out}} = \frac{46.5 \ g}{44.6 \ cc} = 1.04 \text{ g/cc}$$

Percentage of volume under the given saltwater

$$x = \frac{\rho_{eff}}{\rho_0} = \frac{1.043 \text{ g/cc}}{1.1 \text{ g/cc}} = 0.95.$$

This says that 95% of the steel ball will be submerged.

6.7 SURFACE TENSION

If you place a stainless steel needle carefully upon a calm surface of water, it will float even though the density of steel is considerably more than the density of water and the needle is a compact solid (Fig. 6.21). The needle floats not due to the buoyancy as you can easily demonstrate by placing the needle infinitesimally under the surface in which case the needle will sink. Instead, the needle floats because there is an upward force on the needle from the surface molecules.

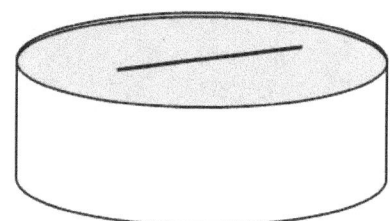

Fig. 6.21: The weight of the floating needle is balanced by the force from the surface tension of water.

The reason for the different behavior at the surface can be found in the special situation of the surface molecules as compared to the molecules in the volume. A molecule inside the volume of the liquid has other molecules of the fluid all around it so that on average there is no net intermolecular force on the molecules in the bulk. But, a molecule on the surface, on the other hand, has molecules of the liquid only on one side, which results in a net intermolecular force pointed inwards from the surface in the liquid (Fig. 6.22). The intermolecular force on a surface molecule is balanced by the net upward impulse from movement of molecules below the surface.

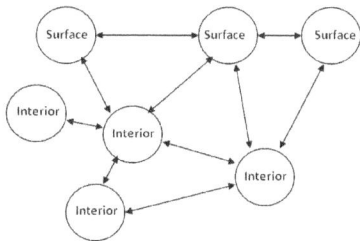

Fig. 6.22: Intermolecular forces on a surface molecule are not balanced.

When we place the needle at the surface, the weight of the needle pushes the surface molecules aside and increases the surface area of the fluid. To expand the surface area of a liquid, requires bringing additional molecules from the inside to the surface against the net attractive force of the other molecules. Therefore, it costs energy to create a larger surface area.

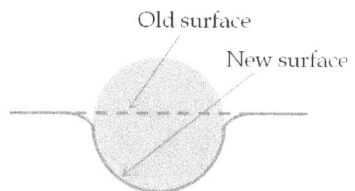

Fig. 6.23: The weight of the floating needle stretches the surface of the fluid, which leads to an increase in upward force from the surface that balances the weight of the needle.

The **surface tension** of a liquid is defined to take into account the energy needed to increase the surface area. Thus, if it takes an energy ΔU to increase the surface area of a fluid by ΔA, then the surface tension (γ) is defined as:

$$\boxed{\gamma = \frac{\Delta U}{\Delta A}} \qquad (6.27)$$

In the absence of external forces, a liquid will tend to assume the least surface area so as to minimize the surface energy. Since a spherical shape has the least area per unit volume, a liquid forms spherical droplets in the absence of external forces such as in the zero gravity situation of outer space (Fig. 6.24).

Fig. 6.24: Marble-sized drop of water filled with tiny bubbles floating in International Space Station. Photo credit: Ken Bowersox, ISS Expedition 6 Commander, NASA, April 2003.

The dimension of surface tension can be obtained from the dimensions of energy and area.

$$[\gamma] = \frac{[M][L]^2/[T]^2}{[L]^2} = \frac{[M]}{[T]^2}.$$

The unit of surface tension in SI system is J/m^2, or equivalently N/m or kg/s^2. More commonly used unit of surface tension, however, is the unit dyne/cm that belongs to the cgs-system. Since 1 N equals 10^5 dyne and 1 m equals 10^2 cm, therefore 1 N/m would equal 10^3 dyne/cm.

$$1\,\frac{N}{m} = 10^3\,\frac{dyne}{cm}.$$

Pure water at 20°C has a surface tension of 72.8 dyne/cm compared to 22.3 dyne/cm for ethyl alcohol, 32 dyne/cm for olive oil, 63.1 dunes/cm for glycerin and 465 dyne/cm for mercury.

Table 6.3: Surface Tension (dyne/cm)
(Divide by 1000 to obtain N/m) (Source: various)

Liquid	Temp (0°C)	Surface Tension γ (dyne/cm)
Water	0	75.6
	20	72.8
	100	58.8
Soapy water	20	~ 40
Blood	37	~ 60
Ethyl alcohol	20	22.4
Glycerin (Glycerol)	20	63.4
Mercury	20	480
Olive oil	20	32.0

The soap lowers the surface tension of water. With the reduced surface tension, the water molecules can squeeze through microscopic spaces between the fibers of the fabric, and remove any trapped dirt particles. The detergent molecules themselves may not be able to pass through the space between fabric fibers, but help water molecules to pass by lowering the surface tension of the mixture.

Similarly, hot water is better at cleaning than the cold water since with rise in temperature the surface tension of water drops, e.g. the surface tension of water is 72.8 dyne/cm at 20°C and 58.9 dynes/cm at 100°C. Table 6.3 provides surface tension of some common substances. *Note that surface tension is not a force but force per unit length* as will be illustrated in the following example.

Example 6.8. Stretching a Planar Soap Film.

Consider a thin film of soapy water produced on a wire frame where one end is movable as shown in Fig. 6.25.

The force needed to move the wire outward by a distance Δx depends on the width w of the film. Furthermore, there are two sides of the film that need to be stretched together. Surface energy will increase due to the work done by the applied force F.

$$\Delta U = F \Delta x \quad \text{(since force } \vec{F} \text{ is parallel to displacement.)}$$

The film has two surfaces, one top and the other bottom. Therefore, the amount of increase in area would be twice the area on one side.

$$\Delta A = 2 \times w \Delta x \quad \text{(two surfaces, each stretching } \Delta x.)$$

Therefore, the surface tension will be

$$\gamma = \frac{\Delta U}{\Delta A} = \frac{F}{2w}.$$

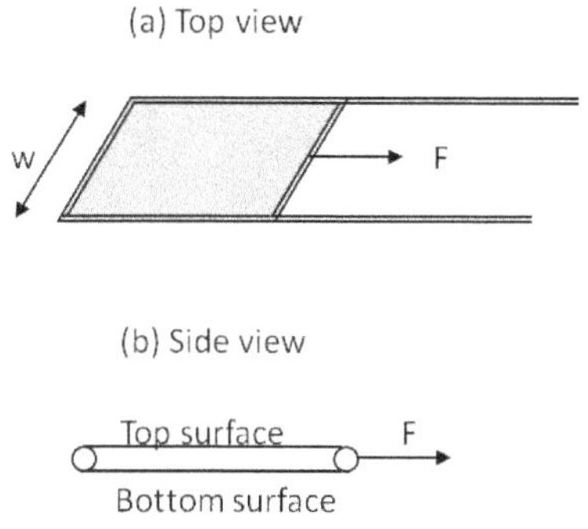

(a) Top view

(b) Side view

Fig. 6.25: Stretching of a soap film.

Example 6.9. Surface Energy of a Water Droplet

Assuming surface tension to be independent of the size of the bubble, find the energy stored in the surface energy of a spherical water droplet of 1-mm radius at $20°C$.

Integrating the equation $dU/dA = \gamma$ from $r = 0$ to $r = R$, we find that the surface energy is equal to the product of surface tension and surface area of the spherical droplet.

$$U(R) = 4\pi\gamma R^2.$$

Note that in the droplet there is only one surface as opposed to the soap film on a wire frame where there were two surfaces. We have used $U(0) = 0$ since if there is no droplet there will be no surface energy. Putting the numerical values we obtain.

$$U(R) = 4\pi \times (72.8 \times 10^{-3} N/m) \times (10^{-3}m)^2 = 9.1 \times 10^{-7} J.$$

Capillary action

Recall from our discussion above that a liquid has the same level in all open tubes connected by freely flowing pathways. As a result when you dip a wide tube in a fluid, the level of the fluid in the tube is same inside the tube as outside. However, when very thin tubes are dipped in a liquid, this is not the case. Instead, liquid rises to various heights depending on the diameter of the opening as illustrated in Fig. 6.26. The effect is called the capillary action.

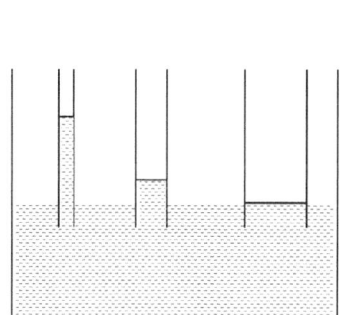

Fig. 6.26: Illustration of the capillary action.

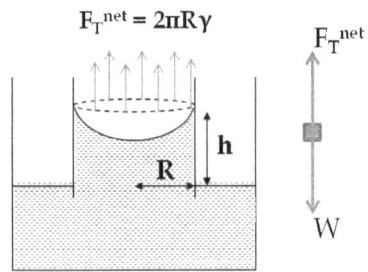

$F_T{}^{net} = 2\pi R\gamma$

Fig. 6.27: Balancing forces on the fluid column.

Let us consider an open tube of radius R dipped in a fluid of density ρ. The height h to which the fluid will rise in the tube can be found by balancing forces on the column of the fluid in the capillary shown in Fig. 6.27.

The capillary action is due to the adhesion of liquid molecules to the tube material and the surface tension. Due to the surface tension along the surface, the surface molecules of the fluid pull on the wall, which in turn, pulls on the fluid upward. The pull of the wall, called adhesive force, supports the weight of the fluid. Therefore, the net upward force from the adhesion force has the magnitude:

$$F_T^{net} = 2\pi R\gamma$$

where γ is the surface tension of the liquid, which must balance the weight of the liquid in the tube given by

$$W = \pi R^2 h \rho g.$$

Equating the two we obtain the height the fluid will rise in a capillary as

$$h = \frac{2\gamma}{\rho g R}.$$

We find that height h is inversely proportional to the tube radius R. If we use the surface tension of water at 20°C in this relation, we find that water will rise 3 cm in a tube of inner diameter 1 mm.

6.8 EXERCISES

Definition of Pressure

Ex 6.1. 1. In a zero gravity environment, some gas is kept at a pressure of 50000 Pa in a rigid cylinder with movable piston. The area of the movable piston is 20 cm^2. What force must be applied on the piston so that it does not move?

Ex 6.2. In a low pressure laboratory, partial vacuum of 10^{-6} Torr is achieved. (a) What is the corresponding reading in millibar? (b) How much is that in psi? (c) How much is that in Pa?

Ex 6.3. Compare the pressures by a 1 N force on a pin and a nail. The tip of the pin has a diameter of 10 μm and the nail 100 μm.

Ex 6.4. On a windy day, the air pressure outside a 90 cm x 200 cm door is 0.8 atm, while the air pressure inside the building is 1.0 atm. How much force must be applied at the handle of the door to open it if the door is hinged on the other end of the 200-cm side?

Ex 6.5. The fuselage of Boeing 777 aircraft is approximately a cylinder of diameter 6 m and length 64 m. If the pressure of the interior of the aircraft is maintained at 0.8 atm, what will be the force tearing the two halves of the aircraft lengthwise when the aircraft is flying at an altitude of 40,000 feet where the pressure outside is 0.2 atm? Hint: Horizontal components of forces cancel out.

Variation of Pressure

Ex 6.6. Depth in a pond varies from 1 meter to 5 meters. What is the range of pressure at the bottom across the pool?

Ex 6.7. Acceleration due to gravity at moon is $\frac{g}{6}$. What is the pressure at a depth of 2 meters in a tank of water at moon?

Ex 6.8. Deduce a formula for the rate of change of pressure in a fish tank in an accelerating elevator when the elevator is (a) going up at acceleration $\vec{a}$ pointed up, (b) going down with acceleration a pointed up, (c) going down at acceleration a pointed up, and (d) going down at acceleration a pointed down. Hint: Direction of motion is not important.

Ex 6.9. Plot $p(y)$ vs y for atmosphere up to a height of 20,000 meters.

Ex 6.10. What is the air pressure at 40,000 feet?

Ex 6.11. What is the difference of air pressure at the top and bottom of a 10-m tall sealed container when the bottom of the container is 100 m from the ground?

Ex 6.12. Solve $dp/dy = -3p^2$, given $p(0) = 1$ atm.

Ex 6.13. Solve $dp/dy = -3\exp(-2y)$, given $p(0) = 1$ atm.

U-tube

Ex 6.14. An open U-tube has uniform cross-sectional area of 3 square cm and same height on both ends. Mercury is poured in it till the height from the top is 50-cm. How much water must be poured on one side so that the level of the mercury rises by 2 cm on the other side?

Ex 6.15. Ethyl alcohol is now poured on the other side of the mercury in exercise 1. How much ethyl alcohol must be poured so that the water level on the other side rises by 1 cm?

Ex 6.16. An open U-tube has different cross-sections on the two sides. On one side the cross-section is 2.5 sq. cm. and on the other side it 4.0 sq. cm. Water is poured so that the level is 30 cm from the top on both sides. How much oil of density 0.8 g/cc must be poured on the wider end so that water rises by 2 cm on the other end?

Pascal's Principles

Ex 6.17. A freely movable piston of mass 100 kg covers water in a rigid cylinder of radius 5 cm touching the surface of the water. (a) What is the pressure at a depth of 50 cm from the top of water? A block of steel of mass 300 kg is placed on the piston. (b) What is the new pressure at a depth of 50 cm from the top?

Ex 6.18. In a hydraulic jack, a 4000 kg pickup truck is supported on the side with cross-sectional area 0.3 m^2 by a 100-N force act on the narrower end. (a) Find the area of cross-section of the narrower end. (b) A refrigerator weighing 3000 kg is loaded on the truck. Now, what force must be used to support the truck?

Ex 6.19. A cylindrical tank containing oil is under pressure such that at the bottom pressure is 3 atm, while at the top the pressure is 1.5 atm. The cover of the container is removed so that the pressure at the top is now 1 atm. What is the pressure at the bottom now?

Archimedes's Principle

Ex 6.20. A plastic spherical ball of negligible mass filled with air has a radius of 10 cm. A boy pushes the ball in water till quarter of its volume is submerged. How much force the boy must apply to keep the ball so submerged? Assume negligible change in the volume of the ball.

Ex 6.21. A 1-mm thick steel sheet is made into a cylindrical cup of area of radius 10 cm and height 30 cm with one end sealed and the other end open. It is then placed in water with the closed end first. How much of the cup stays above water?

Ex 6.22. A cubic wooden block (density 0.9) of side 5 cm is placed in water. (a) How much of the block is under water? (b) The block is then pushed down by a force which balances the increase in the buoyancy force at each instant. What is the total work needed to fully submerge the block?

Ex 6.23. A closed cubical box of side a made of aluminum sheet floats in water with $\frac{1}{4}$ of it above the surface of water. From the given information and known densities of aluminum and water, determine the thickness t of the sheet as a function of a?

Ex 6.24. A cubical box of side 10 cm made of steel has an open side. With open side of the box up, 90% of the box floats in mercury. How much of the box will float if 200 gram of load is put inside the box?

Ex 6.25. A block of brass of mass 800 grams is attached to a string and hung from a support. The block is then placed so that half of the block is submerged in the water. What is the tension in the string?

Ex 6.26. The brass/string/water together with the support and the container of 6.25 are placed in an elevator. (a) What is the tension in the string when the accelerator moves up with an acceleration of 3 m/s^2 pointed up? (b) What is the tension in the string when the elevator moves down with an acceleration of 3 m/s^2 pointed up?

Surface Tension

Ex 6.27. A rectangular film of mercury has a length of 2 cm and width 1 cm. It is stretched by 1.5 mm along its length while keeping the width same. What is the increase in surface energy of the film?

Ex 6.28. A perfectly spherical bubble of diameter 2 cm is blown with Olive oil aboard the International Space Station. How much work was done in going from 1 cm to 2 cm diameter?

Ex 6.29. An liquid of density 2 g/cc rises by 1.8 cm in a capillary tube of radius 0.075 cm when dipped vertically in the liquid. What is its surface tension?

6.9 PROBLEMS

P 6.1. A U-tube has a uniform cross-sectional area of 3 cm^2. First 2 kg water is poured in, and then, one end is sealed. The closed end has air of volume 60 cm^3. Now, 0.5 kg of mercury is poured in the tube from the open end. Find the rise in the water level in the close end. Assume air in the closed end obeys Boyle's law $p_1 V_1 = p_2 V_2$ where p_1, V_1 and p_2, V_2 are pressures and volumes of air at two instances.

P 6.2. A U-tube open at both ends has water on one side and olive oil on the other. The two liquids meet exactly at the bottom. If the height on the oil side is 10 cm, what is the height on the water side?

P 6.3. A tank is filled with water to $\frac{1}{4}$ of the height, and the rest is filled with oil of density 0.7 g/cc. How does the pressure in the tank vary with depth?

P 6.4. The density of a fluid changes with depth according to the following formula.
$$\rho(h) = \rho_0 + ch,$$
where h is the height from the top, ρ_0 the density at the top, and c a small constant. Find a formula for the variation of pressure with depth.

P 6.5. A tube filled with a fluid of uniform density ρ is put in an elevator. What is the pressure at depth h if the elevator has an upward acceleration of magnitude a?

P 6.6. A steel ball of mass 2 kg is hung from a support, and dipped in water as shown. The arrangement is put in an elevator. (a) What will be the tension in the string when the elevator has an upward accelerating of magnitude 2 m/s^2? (b) What will be the tension in the string when the elevator has a downward acceleration of magnitude 2 m/s^2? (c) What is the tension in the string when the elevator is in free fall?

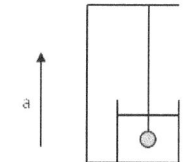

Fig. 6.28: Problem 6.6.

P 6.7. A piece of plastic floats in water with 90% of its volume submerged. An oil of density 0.7 g/cc is poured over the water. It is found that less volume of plastic piece is submerged in water. Find the fraction of volume of the plastic piece under water when it is completely submerged in oil.

P 6.8. One way of finding surface tension of a liquid is based on the measurement of the magnitude of the force F needed to lift a ring from the surface of a liquid. Let R be the radius of the ring. Derive a formula for the surface tension γ in terms of F and R.

P 6.9. The pressure inside a soap bubble is higher than the pressure outside. Derive a formula for the difference in the pressure in terms of the surface tension γ and the radius R of the bubble.

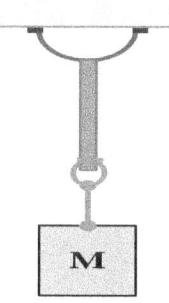

P 6.10. How much weight can be supported by a 4-in diameter suction cup stuck to a flat glass ceiling in such a way that there is a perfect vacuum inside the suction cup.

Fig. 6.29: Problem 6.10.

Chapter 7

FLUID DYNAMICS

Contents

In the last chapter we studied properties of fluids at rest. We found that pressure in the fluid varies with depth, and that Pascal's principles help us understand the distribution of pressure in the fluid. When a fluid is flowing, we find that the pressure at a point in the fluid depends upon the speed of flow also. To distinguish the two pressures, the pressure in the static situation is often called the **static pressure**, while the pressure in a flowing fluid is called the **dynamic pressure**. So, a word of caution at the start here: you should be aware that, the often-used formula $(p - p_0 = \rho g h)$ of the last chapter, is no longer valid when a fluid has a significant flow speed. In this chapter we will find the correct relation between the pressure and the motion of the fluid that replaces the static formula.

Fluid dynamics, or hydrodynamics when the fluid is water, helps us understand a variety of physical phenomena, for instance, the flow of water in river and water pipes, wind around the wings of air planes, the flow of blood in the body, just to name a few. Even with such a wide variety of fluid situations, there are basically two types of fluid flow, called, laminar flow and turbulent flow (Fig. 7.1).

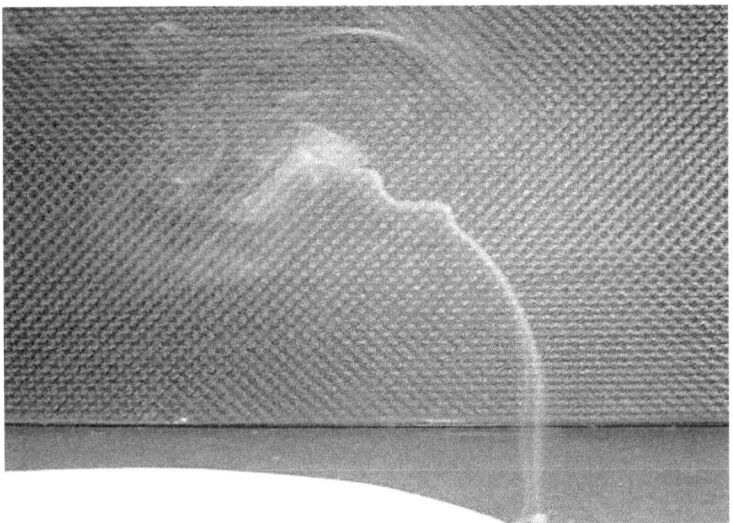

Fig. 7.1: Laminar and chaotic flows can be seen in the smoke of burnt paper. Paper was folded and burnt at the end and a picture of the flow was taken against a dark background.

In the laminar (meaning layered) flow or stream, the fluid moves smoothly in layers with each particle moving in a smooth streamline so that the streamlines of different particles do not cross each other. In the turbulent flow, we find vortices, swirls and chaotic behavior. Turbulence is a very complicated phenomenon and is an active area of research. In this chapter we will study mostly the steady laminar flow and will say only a little about turbulence at the end.

7.1 MASS CONSERVATION AND STEADY FLUID FLOW

The rate of flow of a fluid is described by the current or mass rate of flow, which is the rate at which mass of the fluid moves past a point.

$$\boxed{\text{Mass rate of flow, Current} = \frac{dm}{dt}} \tag{7.1}$$

To be concrete, consider a uniform flow through a pipe of cross-section area A with velocity $\vec{v}$. In duration Δt, you will find that the fluid within a distance $v\Delta t$ would pass through the cross-section of area A (Fig. 7.2). If the density of the fluid is ρ, then the mass of fluid flowing through A per unit time would be equal to $[\rho \times (Av\Delta t)]$. Therefore, the **mass rate of flow**, also called **current**, is

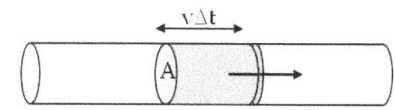

Fig. 7.2: Volume of the liquid that will cross the area of cross-section in unit time.

$$\text{Mass rate of flow, Current} = \rho A v. \tag{7.2}$$

The volume rate of flow is obtained by dividing both sides by the density.

$$\boxed{\text{Volume rate of flow} = Av.} \tag{7.3}$$

In a steadily flowing fluid, fluid does not accumulate at any point. Therefore the density does not change in time. We can use this observation about a steadily flowing fluid to relate the flow properties at two different points. Towards that end, let us consider a steadily flowing fluid in a pipe that varies in the cross-section as shown in Fig. 7.3. Let A_1 and A_2 be the cross-sectional areas of the entrance and exit respectively, and v_1 and v_2 be the corresponding speeds. Let the density of fluid at the entrance and exit be ρ_1 and ρ_2 respectively.

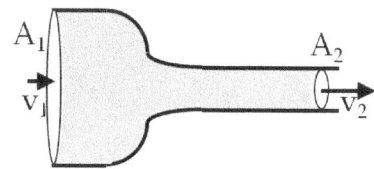

Equating the mass of the material entering the space between A_1 and A_2 in an arbitrary time interval Δt with the amount exiting.

$$A_1 v_1 \rho_1 \Delta t = A_2 v_2 \rho_2 \Delta t.$$

Fig. 7.3: Geometry for derivation of the equation of continuity. The amount of liquid passing 1 must equal the amount leaving 2 if the liquid is incompressible.

Therefore in a steady flow

$$\boxed{A_1 v_1 \rho_1 = A_2 v_2 \rho_2.} \tag{7.4}$$

Equation 7.4 is called the equation of continuity. The continuity equation, as you have seen above, is just a statement about the conservation of mass of the fluid. For a uniform density fluid, the constant density cancels out from the two sides and we obtain a simpler equation.

$$\boxed{A_1 v_1 = A_2 v_2 \ \text{(uniform density).}} \tag{7.5}$$

The area of cross-section times the velocity of flow is called volume rate of flow. Thus we find that in a steady flow of uniform density fluid, the volume flow rate is constant.

The equation of continuity, either Eq. 7.4 or 7.5, together with the Bernoulli's equation, to be derived in the next section, is sufficient to understand the steady non-viscous flow of fluids.

Further Remarks: The Continuity Equation - General

Although we used the entire volume between entrance and exit of a pipe carrying a fluid, the equation of continuity holds for any fluid volume element whatsoever. To derive a more general equation, consider a small rectangular volume between (x, y, z) and $(x+\Delta x, y+\Delta y, z + \Delta z)$ inside the fluid.

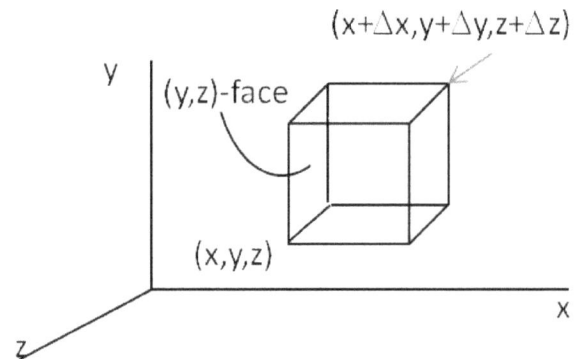

Fig. 7.4: Set-up for calculation of general continuity relation.

Let $\vec{v}(x, y, z)$ be the velocity at the (y, z)-face at (x, y, z) and $\vec{v}(x+\Delta x, y, z)$ be the velocity at the (y, z) face at $(x + \Delta x, y, z)$. Then the amount of fluid moving in interval Δt in the volume element through (y, z) face of the cube at (x, y, z) is $\rho v_x(x, y, z)\Delta y \Delta z \Delta t$, while the amount leaving the (y, z) face at $(x + \Delta x, y, z)$ is $\rho v_x(x + \Delta x, y, z)\Delta y \Delta z \Delta t$. Similarly for the (x, y) and (z, x) faces of the cube shown in Fig. 7.4. Hence, the net change of material inside the volume $\Delta x \Delta y \Delta z$ is:

$$\Delta m = \Delta t \begin{bmatrix} \{\rho v_x(x, y, z) - \rho v_x(x + \Delta x, y, z)\} \, \Delta y \Delta z + \\ \{\rho v_y(x, y, z) - \rho v_y(x, y + \Delta y, z)\} \, \Delta z \Delta x + \\ \{\rho v_x(x, y, z) - \rho v_x(x, y, z + \Delta z)\} \, \Delta x \Delta y \end{bmatrix}$$

Now, we divide both sides by the volume $\Delta x \Delta y \Delta z$ of the volume element at (x, y, z) and by the time interval Δt, and we take the $\Delta t \to 0$ and $\Delta x \Delta y \Delta z \to 0$ limits to obtain the general statement of continuity equation.

$$\frac{\partial \rho}{\partial t} = -\left[\frac{\partial}{\partial x}\left(\rho v_x\right) + \frac{\partial}{\partial y}\left(\rho v_y\right) + \frac{\partial}{\partial z}\left(\rho v_z\right)\right], \qquad (7.6)$$

where I have replaced $\Delta m/\Delta x \Delta y \Delta z$ by $\Delta \rho$ and written $\Delta \rho/\Delta t$ as $\partial \rho/\partial t$. The continuity equation will hold at every point in the fluid as long as matter is neither created nor lost at that point. By introducing a vector current density $\vec{J}$ as

$$\vec{J} = \rho \vec{v}, \qquad (7.7)$$

we can write the continuity equation more compactly as

$$\frac{\partial \rho}{\partial t} = -\vec{\nabla} \cdot \vec{J}, \qquad (7.8)$$

where $\vec{\nabla} = \frac{\partial}{\partial x}\hat{u}_x + \frac{\partial}{\partial y}\hat{u}_y + \frac{\partial}{\partial z}\hat{u}_z$.

7.2 ENERGY CONSERVATION AND BERNOULLI'S EQUATION

The application of the principle of the conservation of energy to the non-viscous steady flow leads to a very useful relation between the pressure and the flow speed in the fluid. This equation is called Bernoulli's equation after Daniel Bernoulli (1700-1782) who published his studies on fluid motion in the book, *Hydrodynamica* (1738).

Consider a fluid flowing through a pipe that may have different cross-sections and be located at different heights at along the flow of a fluid as shown in Fig. 7.5. Let us focus on two such places marked 1

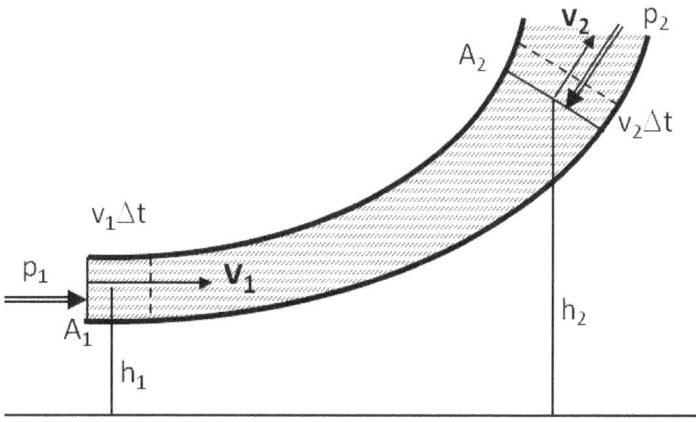

Fig. 7.5: Geometry used for derivation of the Bernoulli's equation.

and 2, and denote the area of cross-section A, the speed of flow v, the height from ground h, and the pressure p at these sites by subscripts 1 and 2 respectively. The density at the two sites will be taken to be the same and denoted by ρ without any subscripts.

We will assume that there are no viscous forces in the fluid so that the energy of any part of the fluid would be conserved. To deduce the Bernoulli's equation, we will equate the change in the energy of the fluid between the points 1 and 2 to the work by the fluid outside of 1 and 2 during an interval Δt. That is, we select the fluid between 1 and 2 as our system and the rest of the fluid and gravity become external to the system.

At point 1, the fluid external to the system is pushing on the system with a force $p_1 A_1$ and in the time interval Δt the boundary moves a distance $v_1 \Delta t$. Therefore, the work done on the system at point 1 is

$$W_1 = (p_1 A_1)(v_1 \Delta t). \tag{7.9}$$

On the other end, the system pushes against the external fluid with a force of magnitude $p_2 A_2$, which must push the system back with a force of the same magnitude but in the opposite direction. Since the displacement of the layer at point 2 during the time interval Δt is $v_2 \Delta t$, the work done on the system at point 2, that is W_2 will be

$$W_2 = -(p_2 A_2)(v_2 \Delta t). \tag{7.10}$$

Therefore the net work done on the system by the external fluid is the sum of W_1 and W_2.

$$W = W_1 + W_2 = p_1 A_1 v_1 \Delta t - p_2 A_2 v_2 \Delta t. \tag{7.11}$$

During the interval Δt, some fluid enters the system at height h_1 with speed v_1 while the same amount leaves the system at another height h_2 with speed v_2. Therefore there will be the following changes in the kinetic energy ΔK and the potential energy ΔU.

$$\Delta K = \frac{1}{2} m_2 v_2^2 - \frac{1}{2} m_1 v_1^2 = \frac{1}{2} (\rho A_2 v_2 \Delta t) v_2^2 - \frac{1}{2} (\rho A_1 v_1 \Delta t) v_1^2 \tag{7.12}$$

$$\Delta U = m_2 g h_2 - m_1 g h_1 = (\rho A_2 v_2 \Delta t) g h_2 - (\rho A_1 v_1 \Delta t) g h_1 \tag{7.13}$$

The work-energy theorem relates the change in the mechanical energy of a system to the work done on the system by external forces. Therefore, we obtain the following relation between the net external work and the change in mechanical energy.

$$W = \Delta K + \Delta U. \tag{7.14}$$

Substituting for W, ΔK and ΔU, rearranging terms and using the equation of continuity $A_1 v_1 = A_2 v_2$ applicable here, we find the following equation, called the Bernoulli's equation.

$$p_1 + \frac{1}{2}\rho v_1^2 + \rho g h_1 = p_2 + \frac{1}{2}\rho v_2^2 + \rho g h_2. \qquad (7.15)$$

This relation states that the mechanical energy of any part of the fluid changes as a result of the work done by the fluid external to that part due to varying pressure along the way. Since the two points were chosen arbitrarily, we can write the Bernoulli's equation more generally as a conservation principle along the flow.

$$p + \frac{1}{2}\rho v^2 + \rho g h = \text{constant.} \qquad (7.16)$$

The constancy of the combination of pressure and the sum of kinetic and potential energy densities is not only constant over time but is also constant over space. A special note must be made here of the fact the in a dynamic situation, the pressure at the same height in different parts of the fluid may be different if they have different speeds of flow.

Special cases of interest:

1. Static situations

Bernoulli's equation includes static behavior as we can see immediately by setting $v_1 = v_2 = 0$ in the equation.

$$p_1 + \rho g h_1 = p_2 + \rho g h_2. \quad \text{(Static)} \qquad (7.17)$$

This says that pressure difference between two points in a fluid at rest is directly proportional to the height difference as we have studied in the last chapter.

$$p_1 - p_1 = \rho g (h_2 - h_1). \quad \text{(Static)} \qquad (7.18)$$

2. Horizontally flowing fluids

For horizontally flowing case $h_1 = h_2$ Bernoulli's equation simplifies.

$$p_1 + \frac{1}{2}\rho v_1^2 = p_2 + \frac{1}{2}\rho v_2^2. \qquad (7.19)$$

An important interpretation of this equation is that in a fluid, pressure is lower where the speed is higher and vice versa. This conclusion is quite contrary to the "common sense". It would seem that pressure should be greater where the fluid is flowing at a higher speed than where it is flowing at a lower speed, but Bernoulli's equation tells us that just the opposite is true.

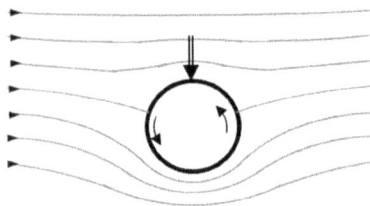

Fig. 7.6: Downward force on spinning curve baseball arises due to increased speed of the air flow at the bottom compared to the speed above the ball when the ball is spinning backwards.

To make use of this aspect of the flow of fluids, airplane wings are constructed so that air will flow with a greater speed above the wings than below the wings. This flow pattern gives rise to a lower pressure at the top of the wing than at the bottom. Therefore, the force pushing on the wing from below would be larger than the force pushing down on the wing from the above. This difference in the force is called the **lift force** and is responsible for flying. (Fig. 7.7).

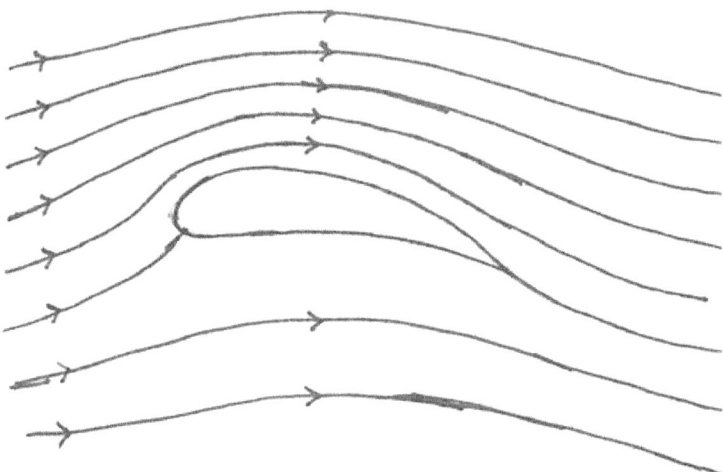

Fig. 7.7: Air flows faster above the wing of an airplane than below the wing aided by the design of the foil. Therefore, pressure above the wing is less than below the wing. The product of the area of the wing and the difference in pressure is the lift force on the wing.

Similarly, in atmospheric physics, we find that the wind speed is higher where the pressure is lower, and in circulatory system, blood flows fastest where the pressure is the lowest.

To understand these and other fluid flow problems, it is usually very helpful to apply Bernoulli's equation to standard problems. To illustrate the applications of the Bernoulli's equation, we will start with the problem of draining a tank that Bernoulli presented in his book Hydrodynamica.

Example 7.1. Draining a tank.

Consider a tank of water with a tap at depth h from the surface in a container. As the tap is opened what is the speed with which water will come out? Assume steady flow. Also assume that the speed of flow of water at the surface in the tank to be very nearly zero compared to the speed at the tap.

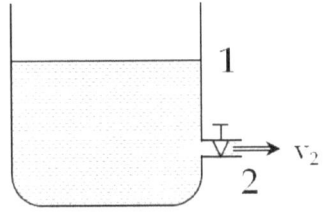

1: $v_1 \approx 0$; $p_1 = p_0$; $h_1 = h$

2: v_2; $p_2 = p_0$ since open to atmosphere; $h_2 = 0$.

Fig. 7.8: Example 7.1

Solution. To apply Bernoulli's equation, we consider two points in the fluid. Point 1 is at the surface and point 2 is just outside the valve at the spigot as shown in Fig. 7.8. Since the area of cross-section at the surface is much larger than the opening, the speed of flow at the surface is much smaller than the speed at the opening. Therefore, I will assume that the speed at a point at the top of the tank may be set to zero without incurring significant error. The pressure at the surface is the atmospheric pressure p_0 since it is freely exposed to the atmosphere.

The pressure against which the water comes out at the opening is also the atmospheric pressure p_0 and not $p_0 + \rho gh$ as would be the case if water was not flowing! This aspect is somewhat disturbing to people first learning to think in terms of dynamics rather than static situation, especially after memorizing formulas of static fluid in the last chapter. The reason for the pressure at the exit to be equal to the atmospheric pressure has to do with the fact that the only the force due to the atmospheric pressure is opposing the flow coming out of the nozzle.

One way to figure out the dynamic pressure at the exit is to pretend that you are the fluid at the exit, and ask, "what force will you be pushing against?" When you think in this language, you would not be confused about the dynamic pressure at any point.

We now put the properties at the two points 1 and 2 in the Bernoulli's equation.

$$p_0 + 0 + \rho gh = p_0 + \frac{1}{2}\rho v_2^2 + 0. \qquad (7.20)$$

Therefore, the speed of water at the spigot is

$$v_2 = \sqrt{2gh} \qquad (7.21)$$

This formula is also called the **Torricelli's theorem**. It is interesting to note that the speed of a freely falling particle dropped from a height h is given by the same formula.

Example 7.2. The Venturi meter

The Venturi meter is a device for measuring the flow speed of gases and liquids of a known density ρ. The fluid at some unknown speed enters at the inlet, passes through a constriction called the throat where a manometer is attached which contains a denser liquid of density ρ_0. The areas of cross-section at the inlet and constriction are A_1 and A_2 respectively.

Fluid ρ: *Applying Bernoulli's equation to the gas at the inlet and outlet at the same height we obtain the following equation.*

$$p_1 + \frac{1}{2}\rho v_1^2 = p_2 + \frac{1}{2}\rho v_2^2 \quad \textit{(dynamic fluid ρ)} \qquad (7.22)$$

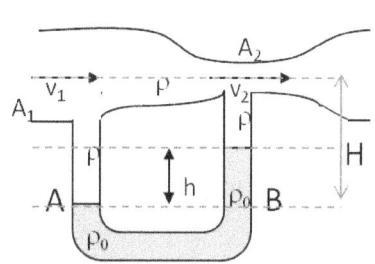

Fig. 7.9: Example 7.2

Fluid ρ_0: *Another relation between the pressures at the inlet and outlet is obtained by applying the static pressure condition on the fluid in the tube. As shown in Fig. 7.9, the pressure at points A and*

B will be equal.

$$p_A = p_B \quad (\text{static fluid } \rho_0)$$

Now, pressure p_A is equal to $p_1 + \rho g H$ and p_B is equal to $p_2 + \rho g (H - h) + \rho_0 g h$. Therefore,

$$p_1 + \rho g H = p_2 + \rho g (H - h) + \rho_0 g h. \tag{7.23}$$

Subtracting Eq. 7.23 from Eq. 7.22 we find

$$(\rho_0 - \rho) g h = \frac{1}{2} \rho v_2^2 - \frac{1}{2} \rho v_1^2. \tag{7.24}$$

The equation of continuity for flowing gas between the inlet and outlet gives us the other relation we need to find v_1.

$$A_1 v_1 = A_2 v_2 \tag{7.25}$$

Eliminating v_2 from Eqs. 7.24 and 7.25, we find the following result for the speed of gas at the inlet.

$$v_1 = A_1 \sqrt{\frac{2(\rho_0 - \rho) g h}{\rho \left(A_1^2 - A_2^2 \right)}}. \tag{7.26}$$

Therefore, $v_1 = \alpha \sqrt{h}$, where α is a constant. Recording the difference in the two arms of the U-tube for the static fluid of density ρ_0 gives us h which lets us determine the flow speed of the flowing fluid of density ρ at the inlet by Eq. 7.26.

Example 7.3. The lift of an airplane

An air plane of mass 1000 kg has air of density 1.2 kg/m^3 flowing above the wing of area 15 m^2 at a speed of 60 m/s over the top surface and 50 m/s past the bottom surface. (a) Find the lift force on the airplane. (b) How does the lift force compare with the weight of the airplane?

Solution. Since the height between the top and bottom of the wing is negligible for the present purpose, we can find the pressure difference between the top and bottom of the wing by using simplified Bernoulli's equation.

$$p_1 + \frac{1}{2} \rho v_1^2 = p_2 + \frac{1}{2} \rho v_2^2$$

$$p_{\text{bottom}} - p_{\text{top}} = \frac{1}{2} \rho_{\text{air}} \left(v_{\text{top}}^2 - v_{\text{bottom}}^2 \right) = 660 \text{ Pa}.$$

The magnitude of the lift force F_L is the pressure difference times area of the wing.

$$F_L = 660 \text{ Pa} \times 15 \text{ m}^2 = 9,900 \text{ N}.$$

(b) The weight of the plane is $W = 1000$ kg $\times$ 9.81 m/s^2 = 9800N. Hence, in the present case, the lift force is 100 N more than the weight.

7.3 VISCOSITY

Real liquids slow down in their flow unless some energy is continuously pumped in the liquid. The intermolecular or **cohesive forces** responsible for the dissipation of energy of flow in liquids are called **viscous forces**.

It is possible to learn about the viscous forces inside the liquid by applying a shear force on the liquid. Unlike solids, fluids do not develop restoring shear force against the external force applied tangentially to the surface. The fluids flow when subjected to a tangential force. However, when you stop applying the external force, the flow slows down due to the work by the internal viscous forces. For instance, when a wind blows over the surface of a lake, the water flows, and when the wind stops, the flow also stops due to the dissipation of energy by way of heat.

There is another way we can access the internal viscous force in a fluid, which is suggested by the wind blowing over the lake. When the wind is blowing over a lake, the speed of flow at the surface is more than the flow deeper down in the water. If you assume a layer-by-layer model of a fluid, then we can study the flow of the fluid layer by layer. We find that the speed of flow of each layer is same. This type of flow is called **laminar flow**. The laminar flow with one surface fixed and the opposite surface subjected to a shear stress will give rise to varying speeds of the layers due to the viscous forces in the liquid as shown in Fig. 7.10.

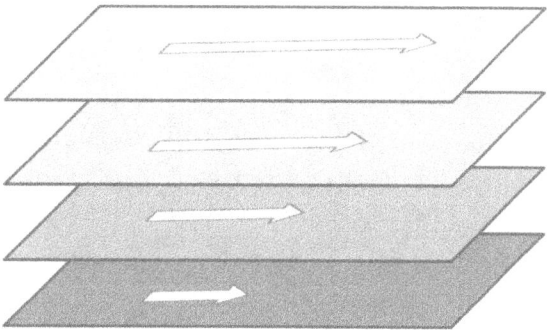

Fig. 7.10: Different layers of a fluid may flow at different speeds. In this figure, the top layer is pushed along by a shear force and the bottom layer is next to a fixed plate.

Now, let us make our understanding a little more quantitative. Consider the following experiment. Place some liquid between two flat plates separated by a distance h. To describe the orientation and directions of vectors, let us introduce a coordinate system. Let the plates be parallel to the xy-plane as shown in the Fig. 7.11. Keeping the bottom plate fixed, let us apply a constant horizontal force F on the top plate to make the top surface move at a steady velocity $\vec{v}$.

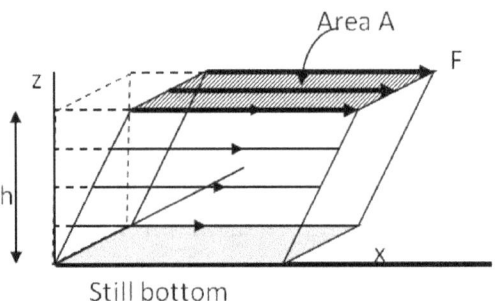

Fig. 7.11: Fluid deformation under a shear stress in a duration Δt. Note the stress force $\vec{F}$ is at the top. The top layer moves a longer distance than all other layers. The layer in contact with the still bottom is not moving.

The liquid layer in contact with the bottom plate will stay at rest while the layer in contact with the top plate will move at the same velocity $\vec{v}$ as for the plate itself. As a first step, and the only step for us, we assume that the velocity of the fluid between the plates increases linearly from zero at the bottom and $\vec{v}$ at the top. This is also true experimentally when the velocity of flow is truly layer-like.

Experiments show that the horizontal force $\vec{F}$ needed to maintain a steady velocity $\vec{v}$ of the top plate is directly proportional to the contact area A and the velocity $\vec{v}$ of the top plate relative to the fixed plate, and inversely proportional to the distance h between the plates as long as the speed v is not too large.

$$\vec{F} = \eta A \frac{\vec{v}}{h} \tag{7.27}$$

where the constant of proportionality η (the Greek letter "eta") is variously called the **coefficient of viscosity, dynamic viscosity,** or simply viscosity. The velocity $\vec{v}$ is the velocity of flow of a layer at distance h from the fixed plate. Right next to the fixed plate, $\vec{v} = 0$. The velocity increases layer by layer, and the resulting flow is laminar.

How does the velocity vary inside?

To study the motion at a point inside the fluid, we examine a thin layer of fluid and apply Eq. 7.27 to this layer. There would be no external shear force inside the liquid. However, the shear will be provided for the flow by changing magnitudes of the viscous force between layers of the fluid since the viscous force is proportional to the velocity of flow.

For analytic treatment, let us choose Cartesian coordinates to discuss the layer-by-layer change in the velocity. Let z-axis be pointed from the fixed plate to the moving plate and x-axis point in the direction of the flow. Let the x-component of the velocity of liquid particles with the z-coordinate z be v_x and those with z-coordinate $z + \Delta z$ be $v_x + \Delta v_x$.

The layer at $z + \Delta z$ has a force $F_x^{above}\hat{u}_x$ from the layer above the $z + \Delta z$ layer and layer at z has a force $F_x^{below}\hat{u}_x$ from the layer below the z layer. The difference of the two forces must be balanced by inter-molecular forces so that the fluid molecules between z and $z+\Delta z$ do not accelerate. We call this force the viscous force or force of viscosity. Let us denote this force by $\vec{F}_{visc}$. In the present experiment, we get the x-component of the the viscous force to be

$$F_x^{visc} = F_x^{above} - F_x^{below} = \eta A \frac{\Delta v_x}{\Delta z} \qquad (7.28)$$

Here the z-coordinate of the layer called "above" is $z + \Delta z + \epsilon$ with $\epsilon > 0$ and the z-coordinate of the layer called "below is $z - \epsilon$. Taking $\Delta z \to 0$ limit we find the following dependence of viscous force $\vec{F}_{visc}$ on the flow velocity $\vec{v}$ and the height z from a static layer.

$$F_x^{visc} = \eta A \frac{dv_x}{dz} \qquad (7.29)$$

Many fluids, such as water, ethyl alcohol, glycerin, and most gases exhibit this simple linear dependence of shear on the gradient of velocity. They are called **Newtonian fluids**. When a fluid does not behave as postulated above, it is called a non-Newtonian fluid.

The viscosity coefficient has dimensions of $[M]/[L][T]$ as determined from the dimensions of other quantities in Eq. 7.27.

$$[\eta] = \frac{[F][h]}{[A][v]} = \frac{[M]}{[L][T]}.$$

The unit of the coefficient of viscosity in SI-system is Pa.s, which is also called poiseuille (PI) after Jean Louis Marie Poiseuille (1797-1869). A more common unit is the unit in the cgs-system of units called **poise**, which is equal to 1 dyne.sec/cm^2. The relation between the two system of units are:

$$1\text{poise} = 0.1 \text{ Pa.s} = 0.1 \text{ PI}$$

Table 7.1: **Viscosity of some common liquids at** $20°C$
(To convert to the SI unit of viscosity Pa.s
divide the numbers in the table by 1000)

Liquid	Viscosity centipoise(cP)
Benzene	0.652
Blood	~ 2.7
Carbon tetrachloride	0.969
Castor oil	984
Diethyl ether	0.233
Ethyl alcohol	1.200
Glycerol	1490
Mercury	1.554
Olive oil	84
Water	1.002

The viscosity of many common fluids are usually tabulated in centipoise as in Table 7.1.

Example 7.4. Flow through a cylinder - Poiseuilli's law

Consider a viscous fluid flow through a cylindrical pipe of radius R placed horizontally. Let there be a difference of pressure p maintained across its ends separated by a distance L. The external force supplied by the difference of pressure at the ends is needed to overcome the effects of drag due to viscosity. The fluid layers of laminar flow in the pipe will be in the shape of concentric cylindrical shells. The fluid at the center moves fastest while the fluid next to the non-moving wall is at rest.

We would like to find the velocity profile, i.e., velocity as a function of the distance from the center for steady state flow. Towards that end consider a cylindrical shell of inner radius r and outer radius $r + \Delta r$ about the central axis. The cylindrical shell is taken to be thin enough that we can assume the fluid in the shell to move with the same speed v. The fluid just outside the shell moves at a lower speed and therefore has a retarding effect on the cylindrical shell under consideration, while the fluid just inside moves at a higher speed, and therefore tends to accelerate the fluid in the shell.

The layer immediately outside of radius $r + \Delta r$ applies a force of magnitude $F_{r+\Delta r}$ on the fluid in the cylindrical shell that attempts to slow the fluid in the shell while the force F_r from layer immediately inside radius r attempts to accelerate the fluid in the shell. In addition to these viscous forces, the force from the pressure difference at the ends attempts to accelerate the flow. In a steady state the three forces must be balanced so that there is no acceleration. Let the z-axis be pointed along the tube. Then, the z-component of the force-balancing equation gives the following.

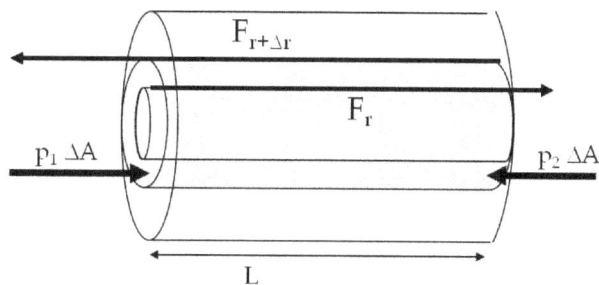

Fig. 7.12: Example 7.4. Forces on the fluid in the cylindrical shell of inner radius r and outer radius $r + \Delta r$. Here the flow is to the right with $p_1 > p_2$. In the text we use Δp for $p_1 - p_2$.

$$p_1 \Delta A - p_2 \Delta A + F_z^r - F_z^{r+\Delta r} = 0.$$

Now, replacing the viscosity forces with their expressions in terms of the coefficient of viscosity and speeds of flow with respect to the wall, where the speed is zero, we obtain

$$\Delta p \times (2\pi r \Delta r) = -\eta 2\pi L \left[\left(r \frac{dv_z}{dr} \right)_{r+\Delta r} - \left(r \frac{dv_z}{dr} \right)_r \right],$$

where we have also replaced $p_1 - p_2$ by Δp. Note that dv_z/dr is negative so we need a minus sign on the right side to give the correct direction of the forces $\vec{F}_{r+\Delta r}$ and $\vec{F}_r$. Dividing both sides by Δr and taking $\Delta r \to 0$ limit, we find the following differential equation.

$$\frac{1}{r} \frac{d}{dr} \left(r \frac{dv_z}{dr} \right) = -\frac{\Delta p}{\eta L} = constant \qquad (7.30)$$

We need to solve this differential equation in order to find the z-component of velocity v_z as a function of distance r from the center. Let us try a polynomial function for v_z as a trial solution.

$$\text{Try: } v_z(r) = ar^2 + br + c.$$

If we find a, b and c then we have the desired answer. Now, since $v_z(r)$ is a maximum for $r = 0$, the coefficient of r^1 must be zero as can be seen by applying elementary result of Calculus: at a maximum or minimum the derivative is zero.

$$\left. \frac{dv_z}{dr} \right|_{r=0} \implies (2ar + b)|_{r=0} = 0 \implies b = 0.$$

Furthermore, at the wall of the pipe, i.e. when $r = R$, the velocity of flow is zero.

$$v_z(R) = aR^2 + c = 0 \implies c = -aR^2.$$

Hence, the velocity has a quadratic profile with a maximum at $r = 0$ and zero at $r = R$ as shown in Fig. 7.13.

$$v_z(r) = a\left(r^2 - R^2\right). \tag{7.31}$$

The remaining undetermined constant a can be found by substituting

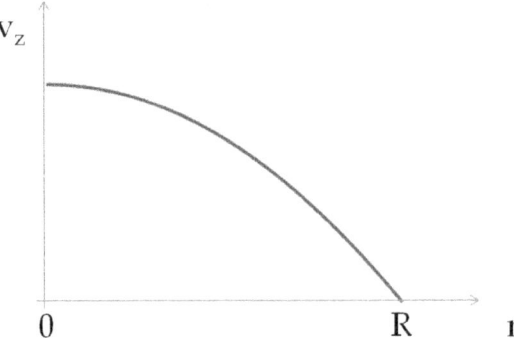

Fig. 7.13: Velocity profile in a tube.

the expression for v_z in Eq. 7.30 giving

$$a = -\frac{\Delta p}{4\eta L}$$

Therefore, the velocity of flow has the following dependence on r.

$$v_z(r) = \frac{\Delta p}{4\eta L}\left(R^2 - r^2\right). \tag{7.32}$$

The volume flow rate through the entire tube can be obtained by summing up the volume flow in each cylindrical shell. The volume rate of flow in the cylindrical shell is equal to the product of its cross-sectional area and the velocity of flow.

$$\text{Volume rate of flow in shell} = \text{velocity} \times \text{cross-sectional area}$$

$$= \frac{\Delta p}{4\eta L}\left(R^2 - r^2\right) \times (2\pi r \Delta r)$$

Summing over infinitesimally thick shells is done by replacing Δr by dr and integrating from $r = 0$ to $r = R$.

$$\frac{d(\text{Volume})}{dt} = \frac{\pi R^4}{8\eta L}\Delta p. \tag{7.33}$$

*This relation is called the **Poiseuille's law**, named after the French physician Jean Poiseuille, who studied fluid flow experimentally in 1840's. Note the very strong dependence of the volume rate of flow on the radius of the pipe; if the radius is doubled then the volume flow rate increases 16-fold! The Poiseuille's law is useful in the study of flow of blood in the body and other engineering applications.*

Example 7.5. Falling sphere in a viscous fluid - Stokes's law.

*Stoke's law for viscous force on a falling spherical ball is an important result that finds use in many areas of physics and engineering. Consider dropping a spherical ball in a still fluid, which can be liquid or gas. As the ball falls, it experiences a drag or resistive force $\vec{F}_{drag}$ depending upon the viscosity of the fluid. Sir George Stokes found the characteristics of the drag force, which is called the **Stokes's Law**. Let R be the radius of the ball, $\vec{v}$ its velocity, and η the viscosity of fluid. Stokes's law states the following for the drag force $\vec{F}_{drag}$.*

$$\vec{F}_{drag} = -6\pi\eta R\vec{v} \tag{7.34}$$

Note that the drag force by Stokes's law is directly proportional to the radius rather than the area of cross-section as one might expect. The formula of the drag force on a sphere can be guessed at by dimensional analysis.

The argument for a dimensional analysis goes as follows. We assume that the magnitude of the drag force will depend on the velocity, $F_{drag} \sim v^m$, for some power m to be determined by the dimensional argument here. We also know that it will involve the size of the ball, although we don't know if it will be R or R^2, or some other power of R, say R^n. The drag force should also be proportional to the viscosity coefficient. Writing out the dimension of the product of these dependencies of the drag force, we find

$$[\eta][v^m][R^n] = \frac{[M][L]^{n+m-1}}{[T]^{m+1}},$$

which should equal the dimension of force, $\frac{[M][L]}{[T]^2}$. Therefore, $m = 1$ and $n = 1$. The dimensional analysis helps us guess the form of the magnitude of the drag force on a spherical particle of radius R up to an undetermined constant C.

$$F_{drag} = C\eta Rv.$$

The calculations that shows that the constant C is equal to 6π is beyond this textbook. The direction of the drag force is opposite to the direction of the velocity of the ball since it is a resistive force for the motion of the ball.

7.4 TURBULENCE

When you let water out of a tap, first you get a smooth laminar flow when the speed of flow is not high, but as you turn up the flow it becomes more roiled and turbulent as shown in a numerical simulation in Fig. 7.14. You see the same transition from a laminar flow to a turbulent flow in all fluids when speed is increased. The mathematics needed to understand the turbulence phenomena is quite complicated. Only recently with the advent of computers has it become possible to find the numerical solutions of some cases of interest such as the flow of air around automobiles and airplanes.

Fig. 7.14: Numerical simulations of supersonic turbulent jet. The simulations were performed by Jonathan Freund, Parviz Moin and Sanjiva K. Lele of Center of Turbulence Research, Stanford University.)

The easiest way to decide whether a flow will be turbulent or laminar is based on a dimensionless parameter called Reynolds number, named after the British engineer Osborne Reynolds (1842-1912) who introduced it in a paper in 1883. Basically, **Reynolds number** (Re) is dimensionless combination of the momentum density, an inertial property, the dissipative property as given by viscosity, and a characteristic length scale of the system, such as diameter of the tube in which the flow occurs.

$$Re = \frac{\rho v D}{\eta} \tag{7.35}$$

where ρ is the density, v the flow speed, η the coefficient of viscosity, and D a characteristic length, e.g. diameter of the cross-section of the tube if the flow is in a cylindrical pipe. Let us check the dimensionless aspect of Re. Putting the units for various quantities, we do find that all the units cancel out.

$$[Re] = \frac{[\rho][v][D]}{[\eta]} = \frac{kg}{m^3} \times \frac{m}{s} \times m \times \frac{m.s}{kg} = \text{Dimensionless}$$

The Reynolds number of a flow is higher for higher speed flow. Reynolds number is also higher for a fluid flowing through a thicker pipe. Experiments show that, regardless of the chemical make-up, the flow is laminar if the Reynolds number for the flow is less than about

2,000, turbulent for Reynolds number above 4000, and in-transition for Reynolds number between 2,000 and 4,000.

If a fluid is less viscous then it will become turbulent at lower speed. Viscosity tends to maintain the layer flow. With increase in speed, the liquid molecules migrate more rapidly from one layer into the other destroying any ordering and the flow develops eddies or vortices at multiple scales. Turbulence mixes fluid rather well. In the internal combustion engine, for instance, air and fuel mix well due to the turbulence in the chamber, which is necessary for a more efficient burn. Turbulence is also an important factor that determines the flight path of baseballs and tennis balls. For a "standard air at sea level", the density and viscosity of air are 1.23 kg/m^3 and 1.73×10^{-3} Pa.s respectively. If a ball of diameter 5 cm is moving at 201 km/h or 126 mph, then from the perspective of the ball the Reynolds number of air will be 2000. Spinning of a baseball gives the air an effective speed greater than the nominal speed, and hence a higher Reynolds. Similarly, the dimples in a golf ball help with the development of turbulence around the ball, which make the drives longer.

Example 7.6. Reynolds number of water

A 2-cm diameter pipe carries pure water of density 1 g/cc and viscosity 1.002 centipoise. At what speed will the Reynolds number be 2000?

Solution. For Reynolds number to be 2000, the water has to flow at 10.02 cm/s obtained directly from the formula. At that flow speed, 113 liters of water will flow per hour. Compare this to the normal kitchen water fountain flow rate in the US which is more than 1000 liters per hour. Clearly when you turn on the kitchen faucet on full, the flow is turbulent. This is visible as frothy and bubbly flow when the water comes out of the faucet.

7.5 EXERCISES

Steady flow

Ex 7.1. Water flows through a cylindrical pipe of internal diameter 1 cm. (a) If 1000 L of water flows per minute, what is the velocity of flow in m/s? (b) What is the mass rate of flow in kg/s?

Ex 7.2. A water supply pipe to a house has an internal diameter of 4 cm. The second floor of the house has water pipe of diameter 2 cm. (a) If water flows at the speed of 10 m/s from the pipe at the second floor, what is the speed of flow in the supply line? (b) What is the volume rate of flow? (c) What does current equal to?

Ex 7.3. Rate of mass flow per unit area is also called current density J. Find the current density of water flowing through a cylindrical pipe of diameter 5 cm at the rate of 100 L per second.

Bernoulli's equation

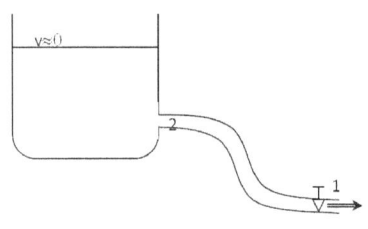

Fig. 7.15: Example 7.4

Ex 7.4. A water tank is drained by a pipe that varies in cross-section, with 5 cm diameter at the tank (point 2) and a diameter of 1 cm at the exit (point 1). Point 2 is 30 cm above point 1. The surface of water in the tank is 50 cm above point 1. (a) Find the speed of flow at point 1. (b) Find volume rate of flow. (c) Find rate of flow at point 2 . (d) Find pressure at point 2.

Ex 7.5. Water is pumped from the ground floor of a high-rise building to a 10th floor apartment 25 meters above. A gauge pressure at the pump reads 400 kPa and the pipe on the 10th floor is open to atmosphere. While the pipe on the ground floor is 5 cm in diameter, that in the apartment is only 2 cm in diameter. Find speeds of flow at (a) the ground floor and (b) the tenth floor.

Ex 7.6. A large water tank is filled with water to the top, and sealed off at the top so that it is not exposed to the atmospheric pressure. A small orifice is opened at the bottom. Determine the minimum water height in the tank for the water to leak at all.

Ex 7.7. On a windy day wind blows on top of a roof at a speed of 110 km/hr. Under the roof air is quiet. If the density of air is 1.3 kg/m^3, what is the lift force per unit m^2 area of the roof.

Ex 7.8. A venturi flow meter is used to measure the flow rate of water by inserting the flow meter between at one place in the pipe where the diameter is 5 cm and another place where the diameter is 1.5 cm. The venturi meter is filled with oil of density 1.8 g/cc. Find the speed of flow of water in the pipe at the 5cm-diameter place if the oil on the two sides of the venturi manometer has a height difference of 2 cm. See Fig. 7.9 for an illustration of a venturi meter.

Viscosity

1. One liter of blood flows per day through a narrow vein of length 5 cm which has a pressure difference of 0.5 atm across its length. What is the diameter of the vein?

2. A spherical steel ball of mass 200 grams is dropped in glycerin (glecerol) from rest. After some time, the ball drops with a steady terminal speed. What is the value of the terminal speed? Ignore the force of buoyancy. (Density of the steel is 7.8 g/cc)

3. A spherical ball of radius R and density ρ is dropped from rest in a fluid of density ρ_0. Derive a formula for the time it takes to reach the terminal speed.

Turbulence

1. Determine which of the following flows will be turbulent. (a) mercury flowing at $20°C$ through a tube of inner radius 1 cm and length of 1 m carrying 30 liters per hour, (b) water flowing in the same tube at the same volume rate, and (c) olive oil flowing in the same tube at the same volume rate.

2. Blood flows through an artery of diameter 3 mm. What is the volume rate of flow if Reynolds number of the flow is 400?

7.6 PROBLEMS

P 7.1. A filled large water tank has a pipe of cross-sectional area 1.2 cm^2 attached at a depth 30 cm from the top. There is a valve controlling the flow of water through the pipe. Find the pressure at

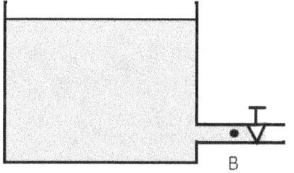

Fig. 7.16: Problem 7.1

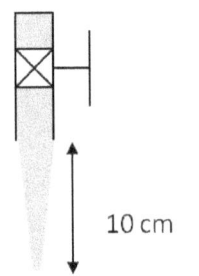

10 cm

Fig. 7.17: Problem 7.2

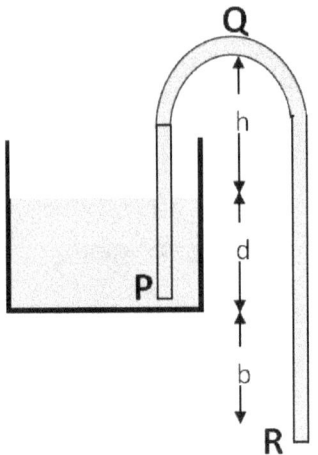

Fig. 7.18: Problem 7.3

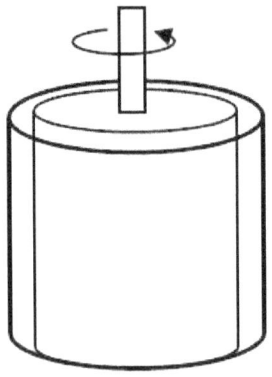

Fig. 7.19: Problem 7.7

a point B in the pipe, (a) when no water is flowing, and, (b) when water is flowing steadily.

P 7.2. As water falls from a faucet it narrows. What will be the area of cross-section of flow 10-cm from the exit of a water faucet that has an opening of 2 cm diameter if 1/4 liter of water flows per second?

P 7.3. A siphon is an ingenious device for removing fluid from a container that must not be tipped. In a siphon, you fill a U-tube or a flexible pipe with a liquid, close both ends, invert the pipe and place the higher end in the liquid and the other end in another container. Now, when you open the ends, you find that fluid starts to drain out as long as the highest point Q of the pipe is not too high. Find the maximum height h so that water can drain when the open end in the water is at a depth of d and the other end R of the siphon is a height b below. Denote the density of the fluid by ρ and give your answer in terms of p_{atm}, ρ, g, d and b.

P 7.4. Refer to the figure in the problem above. If the area of cross-section of the container is A_0, the volume of water in the container is V, and area of cross-section of the pipe is A, how long will it take to empty the container? Do not assume the speed of flow inside the container to be zero.

P 7.5. In an intravenous (IV) line, saline water of density 1.1 g/cc from a raised container flows to a needle with an inside diameter of 0.45 mm, which is inserted in a vein in the arm. Assume the gauge pressure in the vein varies between 20 mmHg and 30 mmHg. Assuming steady non-viscous flow (a) find the range of flow speeds in the needle, and (b) volume rate of flow of the solution per minute, if the container is placed 1.2 meters above the needle?

P 7.6. In a high-rise building, water is pumped from the ground floor to the 30th floor in a pipe of diameter 4 cm. The length of the pipe is 75 meters. What should be the pressure in the pump so that the flow has a Reynolds number of 1000.

P 7.7. A viscometer is a device that measures viscosity of a liquid. In a common viscometer the liquid is put between two drums, one of which is fixed and the other rotated at a constant angular speed. The torque needed to rotate the drum is related to the viscosity. In a particular viscometer the inner drum is rotated at 50 rev/min. The outer diameter of the inner drum is 5 cm and the inner diameter of the outer fixed drum is 6 cm. The height of the drums is 4 cm. If the torque on the inner drum is measured to be 9.75×10^{-4} $N.m$, what is the viscosity of the fluid?

Index Volume 1

Index Volume 2

FUNDAMENTAL PHYSICAL CONSTANTS

http://www.physics.nist.gov/cuu/Constants/index.html

QUANTITY	SYMBOL	VALUE
UNIVERSAL		
Speed of light in vacuum	c	2.99792458×10^8 m/s (Exact)
Magnetic permeability of vacuum	μ_0	$4\pi \times 10^{-7}$ N/A^2 (Exact)
Electric permittivity of vacuum	$\epsilon_0 = 1/c^2\mu_0$	$8.854187817... \times 10^{-12}$ F/m (Exact)
Newtonian Constant	G	$6.6742(10) \times 10^{-11}$ N.m^2/kg^2
Planck Constant	h	$6.6260755(40) \times 10^{-34}$ J.s
h-bar or $\hbar$	$h/2\pi$	$\sim 1.05457 \times 10^{-34}$ J.s
ELECTROMAGNETIC		
Bohr magneton	μ_B	$9.27400949(80) \times 10^{-24}$ J/T
Elementary charge	e	$1.60217653(14) \times 10^{-19}$C
Josephson constant	K_J	$4.83597876(41) \times 10^{14}$ Hz/V
Magnetic quantum flux	Φ_0	$2.06783372(18) \times 10^{-15}$ Wb
Nuclear magneton	μ_N	$5.05078343(43) \times 10^{-27}$ J/T
Von Klitzing constant	R_K	$2.5812807449(86) \times 10^4$ Ω
ATOMIC		
Electron mass	m_e	$9.1093826(16) \times 10^{-31}$ kg
Proton mass	m_p	$1.67262171(29) \times 10^{-27}$ kg
Neutron mass	m_n	$1.67492728(29) \times 10^{-27}$ kg
Bohr radius	a_0	$0.5291772108(18) \times 10^{-10}$ m
Compton wavelength	$\lambda_c = \frac{h}{m_e}$	$2.426310238(16) \times 10^{-12}$ m
Electron gyromagnetic ratio	g_e	$1.76085974(15) \times 10^{11}$ 1/T.s
Proton magnetic moment	μ_p	$1.41060671(12) \times 10^{-26}$ J/T
Rydberg constant	R_H	$1.0973731568525(73) \times 10^7$ 1/m
PHYSICOCHEMICAL		
Atomic mass	u or m_u	$1.66053886(28) \times 10^{-27}$ kg
Avogadro number	N_A	$6.0221415(10) \times 10^{23}$ 1/mol
Boltzman constant	k or k_B	$1.3806505(24) \times 10^{-23}$ J/K
Faraday constant	F	$96485.3383(83)$C/mol
Molar gas constant	R	$8.314472(15)$J/K.mol
Stephan-Boltzman constant	σ	$5.670400(40) \times 10^{-8}$ W/m^2.K^4

USEFUL CONVERSION FACTORS

Length
  1 in = 2.54 cm
  1 ft = 12 in
  1 mi = 5280 ft ≈ 8/5 km
  1 nautical mile ≈ 1.1584 mi ≈ 1.852 km
  1 astronomical unit (au) ≈ 1.5×10^{11} m
  1 light year (ly) ≈ 9.46×10^{15} m
  1 parsec ≈ 3.1×10^{16} m
  1 fermi (fm) = 10^{-15} m
  1 angstrom () = 0.1 nm = 10^{-10} m

Mass
1 pound (lb) = 453.59237 g
1 kg = 0.0685 slug
1 ounce (oz) ≈ 28.35 g
1 (metric) ton = 1000 kg
1 u ≈ 9.1×10^{-31} kg

Volume
1 liter (L) = 1000 cm^3 (cc) = 10^{-3} m^3
1 US-quart (qt) = 32 fluid ounce (oz)= 0.946 L
1 US-gallon (gal) = 4 qt = 0.83 gal (Imperial)

Energy
1 Joule (J) = 1 Volt-Coulomb
= 10^7 ergs
≈ 0.74 lb-force.ft
≈ 9.5×10^{-4} British thermal unit (Btu)
≈ 2.8×10^{-5} kiloWatt-hour (kWh)
1 calorie (cal) ≈ 4.186 J ≈ 4×10^{-3} Btu
1 electron volt (eV) = 1.602×10^{-19} J
1 liter.atm = 101.3 J

METRIC MULTIPLIERS

Factor	Prefix	Symbol	Factor	Prefix	Symbol
10^{1}	deka-	da	10^{-1}	deci-	d
10^{2}	hecto-	h	10^{-2}	centi-	c
10^{3}	kilo-	k	10^{-3}	milli-	m
10^{6}	mega-	M	10^{-6}	micro-	μ
10^{9}	giga-	G	10^{-9}	nano-	n
10^{12}	tera-	T	10^{-12}	pico-	p
10^{15}	peta-	P	10^{-15}	femto-	f
10^{18}	exa-	E	10^{-18}	atto-	a
10^{21}	zetta-	Z	10^{-21}	zepto-	z
10^{24}	yotta-	Y	10^{-24}	yocto-	y

www.ingramcontent.com/pod-product-compliance
Lightning Source LLC
Chambersburg PA
CBHW081436170526
45166CB00008B/2218